J. Paulo Davim
Editor

Nontraditional Machining Processes

Research Advances

Springer

Editor
J. Paulo Davim
Department of Mechanical Engineering
University of Aveiro
Aveiro
Portugal

ISBN 978-1-4471-5178-4   ISBN 978-1-4471-5179-1   (eBook)
DOI 10.1007/978-1-4471-5179-1
Springer London Heidelberg New York Dordrecht

Library of Congress Control Number: 2013940292

Printed on acid-free paper

Springer is part of Springer Science+Business Media (www.springer.com)

Preface

The designation *"nontraditional machining"* refers to a group of processes that removes material by various methods involving electrical, thermal, chemical and mechanical energy (or combinations of these energies). Currently, typical applications of nontraditional machining methods include high accuracies, good surface finish and complex geometries, parts machined without burrs or residual stresses as well as work materials that cannot be machined by conventional methods. In general, the nontraditional processes are characterized by high specific energies and low removal rates when compared to conventional machining processes. Today, nontraditional macro and micromachining processes present great importance to automotive, aircraft, moulds and dies and other advanced industries placed in all industrialized or emerging countries.

Chapter 1 of the book describes *Laser-Assisted Manufacturing: Fundamentals, Current Scenario and Future Applications*. Chapter 2 is dedicated to *Laser Beam Machining*. Chapter 3 describes *Laser Cutting of Triangular Geometry in Aluminum Foam*. Chapter 4 contains information on *Micro-Electrical Discharge Machining*. Chapter 5 describes *Prototype Machine for Micro-EDM*. Chapter 6 contains information on *Abrasive Water Jet Milling*. Finally, Chap.7 is dedicated to *A New Approach for the Production of Blades by Hybrid Processes*.

The present book can be used as a research book for final undergraduate engineering course or as a topic on manufacturing at the postgraduate level. Also, this book can serve as a useful reference for academics, researchers, manufacturing, industrial, materials and mechanical engineers, professionals in nontraditional machining and related industries. The scientific interest in this book is evident for many important centers of research, laboratories and universities as well as industry. Therefore, it is hoped this book will inspire and enthuse others to undertake research in this field of nontraditional machining processes.

The Editor acknowledges Springer for this opportunity and for their enthusiastic and professional support. Finally, I would like to thank all the chapter authors for their availability for this work.

Aveiro, Portugal, February 2013   J. Paulo Davim

Contents

Contributors

S. S. Akhtar Mechanical Engineering Department, King Fahd University of Petroleum and Minerals, Dhahran, Saudi Arabia

P. Bhargava Laser Material Processing Division, Raja Ramanna Centre for Advanced Technology, PO: RRCAT, Indore, Madhya Pradesh 452 013, India

Ivo M. F. Bragança IDMEC, Instituto Superior Técnico, Technical University of Lisbon, Av. Rovisco Pais, Lisbon 1049-001, Portugal

Milan Brandt School of Aerospace, Mechanical and Manufacturing Engineering, RMIT University, Bundoora, VIC 3083, Australia

A. Calleja Department of Mechanical Engineering, Escuela Técnica Superior de Ingeniería Industrial, University of the Basque Country, Alameda de Urquijo s/n, Bilbao 48013, Spain, e-mail: amaia_calleja@ehu.es

A. Fernandez Department of Mechanical Engineering, Escuela Técnica Superior de Ingeniería Industrial, University of the Basque Country, Alameda de Urquijo s/n, Bilbao 48013, Spain

Muhammad P. Jahan Department of Architectural and Manufacturing Sciences, Western Kentucky University, Bowling Green, KY 42101, USA, e-mail: muhammad.jahan@wku.edu

O. Keles Mechanical Engineering Department, Gazi University, Ankara, Turkey

L. M. Kukreja Laser Material Processing Division, Raja Ramanna Centre for Advanced Technology, PO: RRCAT, Indore, Madhya Pradesh 452 013, India

Atul Kumar Laser Material Processing Division, Raja Ramanna Centre for Advanced Technology, PO: RRCAT, Indore, Madhya Pradesh 452 013, India

A. Lamikiz Department of Mechanical Engineering, Escuela Técnica Superior de Ingeniería Industrial, University of the Basque Country, Alameda de Urquijo s/n, Bilbao 48013, Spain

L. N. Lopéz de Lacalle Department of Mechanical Engineering, Escuela Técnica Superior de Ingeniería Industrial, University of the Basque Country, Alameda de Urquijo s/n, Bilbao 48013, Spain

Paulo A. F. Martins IDMEC, Instituto Superior Técnico, Technical University of Lisbon, Av. Rovisco Pais, Lisbon 1049-001, Portugal, e-mail: pmartins@ist.utl.pt

C. P. Paul Laser Material Processing Division, Raja Ramanna Centre for Advanced Technology, PO: RRCAT, Indore, Madhya Pradesh 452 013, India, e-mail: paulcp@rrcat.gov.in

Gabriel R. Ribeiro IDMEC, Instituto Superior Técnico, Technical University of Lisbon, Av. Rovisco Pais, Lisbon 1049-001, Portugal

A. Rodriguez Department of Mechanical Engineering, Escuela Técnica Superior de Ingeniería Industrial, University of the Basque Country, Alameda de Urquijo s/n, Bilbao 48013, Spain

Pedro A. R. Rosa IDMEC, Instituto Superior Técnico, Technical University of Lisbon, Av. Rovisco Pais, Lisbon 1049-001, Portugal

Mukul Shukla Department of Mechanical Engineering Technology, University of Johannesburg, Johannesburg, South Africa ; Department of Mechanical Engineering, MNNIT, Allahabad, Uttar Pradesh, India, e-mail: mshukla@uj.ac.za; mukulshukla@mnnit.ac.in

Shoujin Sun School of Aerospace, Mechanical and Manufacturing Engineering, RMIT University, Bundoora, VIC 3083, Australia, e-mail: shoujin.sun@rmit.edu.au

B. S. Yilbas Mechanical Engineering Department, King Fahd University of Petroleum and Minerals, Dhahran, Saudi Arabia, e-mail: bsyilbas@kfupm.edu.sa

Chapter 1
Laser-Assisted Manufacturing: Fundamentals, Current Scenario, and Future Applications

C. P. Paul, Atul Kumar, P. Bhargava and L. M. Kukreja

Abstract This chapter presents the basic principles, applications, and future prospects of various laser-assisted manufacturing techniques used for material removal, joining, and additive manufacturing. The laser hazard and safety aspect is also briefly included.

1.1 Introduction

The principle of the laser was first known in 1917, when physicist Albert Einstein described the theory of stimulated emission. However, the first laser was practically demonstrated by Theodore Maiman of Hughes Research Laboratories on May 16, 1960 in the form of ruby laser [1]. But this technical break through was dubbed as "solution looking for problems" during early years after the invention. Following the invention of the ruby laser, many other materials were found that could be used as the basis of laser action, such as sapphire, neodymium, and organic dyes such as rhodamine 6G. There were also different ways to excite various compounds to the point of lasing, such as certain chemical reactions, or the acceleration of free electrons to very high energy levels. Today, the laser's presence in the world is ubiquitous [2]. The lasers can heat and vaporize any material, drill holes in the hardest materials—diamond, it can create conditions similar to those on the surface of sun in the laboratory, it can cool the atoms to temperatures almost close to absolute zero, it can measure various parameters with exceptional accuracy, and it can detect impurity. Its continual expansion of the boundaries of science, medicine, industry, and entertainment has resulted in many wonderful applications. Smart bombs, supermarket bar code readers, fiber-optic

C. P. Paul (✉) · A. Kumar · P. Bhargava · L. M. Kukreja
Laser Material Processing Division, Raja Ramanna Centre
for Advanced Technology, PO: RRCAT, Indore, MP 452013, India
e-mail: paulcp@rrcat.gov.in

J. P. Davim (ed.), *Nontraditional Machining Processes*,
DOI: 10.1007/978-1-4471-5179-1_1, © Springer-Verlag London 2013

communication, CD/DVD players, laser printers, certain life-saving cancer treatments, or precise navigation techniques for commercial aircraft could only be possible because of lasers. New and popular medical procedures using lasers have enabled to get rid of eyeglasses, removal of unsightly moles, wrinkles, and tattoos, and even streamline bikini lines. In industries, lasers are employed for a variety of material processing, including—cutting, drilling, welding, brazing, surface hardening, cladding, alloying, and rapid manufacturing. Laser-based manufacturing has several advantages over conventional methods. Some of them are listed below:

1. As non-contact process, it is well suited for processing advanced engineering materials such as brittle materials, electric and non-electric conductors, and soft and thin materials.
2. It is a thermal process and materials with favorable thermal properties can be successfully processed regardless of their mechanical properties.
3. It is a flexible process.

Before proceeding further, first, we will briefly discuss how a laser works, what its special properties are, and how these are being exploited for various applications.

1.2 Lasers Basics

LASER is the acronym of light amplification by stimulated emission of radiation and is essentially a source of intense coherent radiation. The laser differs from ordinary source of light in the emission process of radiation. In an ordinary source, atoms or molecules are excited by thermal excitation, for example, electrical discharge which emits photons spontaneously in a random manner. Photons emitted in all directions with no correlation in wavelength, phase, and polarization between them (Fig. 1.1a). In lasers, photons are emitted by stimulated emission—a characteristic process that generates the photons with all properties (namely— wavelength, phase, direction, and polarization) as those of stimulated photons. Thus, the photons get amplified in an orderly manner (Fig. 1.1b).

Now consider a medium with a large number of atoms (molecules), some of which are in the excited state and rest unexcited. Since the photons get absorbed by

Fig. 1.1 **a** Spontaneous emission. **b** Stimulated emission

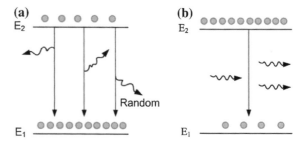

the unexcited atoms, the number of excited atoms should be more than that of unexcited atoms for net amplification. This situation is called population inversion. The name is derived from normal trend of the population in the thermal equilibrium, where unexcited atoms are more abundant. Under normal or thermal equilibrium conditions, the lower energy levels are more highly populated than the higher levels ($N_1 > N_2$), and the distribution is given by Boltzmann's law that relates N_1 and N_2 as

$$\frac{N_2}{N_1} = e^{-\frac{E_2 - E_2}{k_B T}}$$

where k_B is Boltzmann's constant $= 1.38 \times 10^{-23}$ J/K and T is the absolute temperature of the system (K). Population inversion ($N_2 > N_1$) though is the primary condition, but in itself is not sufficient for producing a laser. As there are certain losses of the emitted photons within the material itself in addition to spontaneous emission, one has to think about the geometry that can overcome these losses and there is overall gain.

It derives to the following three prerequisites:

- An active medium with a suitable set of energy levels to support laser action.
- A source of pumping energy in order to establish a population inversion.
- An optical cavity or resonator to introduce optical feedback and so maintain the gain of the system overcoming all losses.

In order to create a laser beam from an active medium, that is, the medium in which population inversion is created, the medium is placed between two mirrors (Fig. 1.2). The photons are reflected back and forth by mirrors and get amplified more and more by the active medium. One of the mirrors is partially transmitting through which the laser beam comes out. Stimulated emission and optical resonator arm the laser with certain unique properties. These properties are briefly discussed in the following sections [3, 4].

1. *Monochromaticity*. The energy of a photon determines its wavelength through the relationship $E = hc/\lambda$, where c is the speed of light, h is Planck's constant,

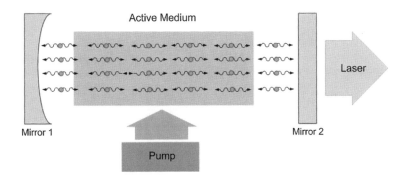

Fig. 1.2 Basic laser system

and λ is the wavelength. In an ideal case, the laser emits all photons with the same energy, thus the same wavelength, and it is said to be monochromatic. However, in all practical cases, the laser light is not truly monochromatic. A truly monochromatic wave requires a wave train of infinite duration. The spectral emission line from which it originates does have a finite width, because of the Doppler effect of the moving atoms or molecules from which it comes. However, compared to the ordinary sources of light, the range of frequency (line width) of the laser is extremely small.

2. *Coherence*. Coherent means that all the individual waves of light are moving precisely together through time and space, that is, they are in phase. Since a common stimulus triggers the emission events, which provide the amplified light, the emitted photons are "in step" and have a definite phase relation to each other. These emitted photons having a definite phase relation to each other generate coherent output, that is, the atoms emit photons in phase with the incoming stimulating photons and emitted waves add to the incoming waves, generating brighter output. Addition is due to the relative phase relationship. Photons of ordinary light also come from atoms, but independent of each other and without any phase relationship with each other and are not coherent. Therefore, laser is called a coherent light source where as an ordinary light is called an incoherent source of light. The concept of coherence can be well understood from the following Fig. 1.3.

There are two types of coherence—spatial and temporal. Correlation between the waves at one place at different times, or along the path of a beam at a single instant, is effectively the same thing and is called "temporal coherence." Correlation between different places (but not along the path) is called "spatial coherence."

3. *Directionality*. One of the important properties of laser is its high directionality. The mirrors placed at opposite ends of a laser cavity enable the beam to travel back and forth in order to gain intensity by the stimulated emission of more photons at the same wavelength, which results in increased amplification due to the longer path length through the medium. The multiple reflections also produce a well-collimated beam, because only photons traveling parallel to the cavity walls will be reflected from both mirrors. If the photon is the slightest bit off axis, it will be lost from the beam. The resonant cavity, thus, makes certain

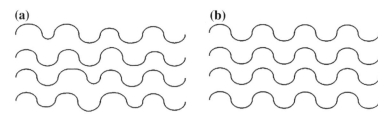

(a)   **(b)**

Fig. 1.3 a Incoherent. **b** Coherent beam

that only electromagnetic waves traveling along the optic axis can be sustained, consequent building of the gain. The high degree of collimation arises from the fact that the cavity of the laser has very nearly parallel front and back mirrors, which constrain the final laser beam to a path, which is perpendicular to those mirrors. For a laser, the beam emerging from the output mirror can be thought of as the opening or aperture, and the diffraction effects on the beam by the mirror will limit the minimum divergence and spot size of the beam (Fig. 1.4).

From diffraction theory, the divergence angle θ_d is:

$$\theta_d = \frac{\beta\lambda}{D}$$

where λ and D are the wavelength and the diameter of the beam, respectively, and β is a coefficient whose value is around unity and depends on the type of light amplitude distribution and the definition of beam diameter. θ_d is called diffraction-limited divergence.

4. *Brightness.* The brightness of a light source is defined as the power emitted per unit surface area per unit solid angle. A laser beam of power P, with a circular beam cross section of diameter D and a divergence angle θ_d and the result emission solid angle is $\pi\theta_d^2$, then the brightness of laser beam is

$$B = \frac{4P}{(\pi D\theta_d)^2}$$

5. *Focussability.* Focusing of laser beams enables high intensity at the focusing spot. Laser power at the spot and its spot size are some of the crucial parameters during various techniques of laser material processing. Very low divergence of the laser beam allows it to be focused to very small spot $\approx \lambda$

6. *High power.* The accumulated photon density directly depends on the density of excited atoms in active medium and volume of active medium. Higher photon density results in higher laser power.

7. *Short-pulse generation.* Very high powers of lasers are also achieved by different methods of pulse generation/pulse compression techniques. Pulsed laser power in µs, ns, and fs time durations are widely available in MWs, GWs, and

Fig. 1.4 Divergence in laser beam

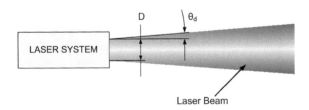

LASER SYSTEM

Laser Beam

TWs power ranges while cw laser power is in the range of 10–100 kW. High harmonic generation allows the generation of attosecond pulses with even higher powers of lasers.

1.3 Lasers and Manufacturing Techniques

Though lasers did wonders for advanced applications of science, engineering, and medicine, but they failed in many simple applications. For example, when it was attempted to cut the chocolate candy and slice the bread, the outcomes were burnt charcoal and toast, respectively. These efforts just failed because of incorrect choice of laser and improper selection of processing parameters. Thus, right choice of laser and proper selection of processing parameters are mandatory to realize the applicability and capability of a laser in any application.

Figure 1.5 presents the schematic arrangement of a typical laser processing station. The system shows a laser integrated with a beam delivery system. The beam delivery system may be optical-fiber based or reflecting optics based. The optical-fiber-based beam delivery system is preferred over reflecting optics based due to ease of operations during the material processing. Various CNC laser workstations are used for the beam or job manipulations during laser material processing. Robots are not very common in laser processing due to inferior position accuracy. Among CNC workstations, 3-axis interpolation (X, Y, and Z) is sufficient to reach to any point in the space, but two more axes (A and C) are required to orient particular direction to reach. Therefore, 3-axis configuration is minimum system requirement and 5-axis is universal requirement without redundancy. Apart from axes-movements, laser workstation needs some more features, like laser on/off, gas on/off, powder feeder on/off, for laser rapid manufacturing (LRM). Depending upon the applications, various processing heads are used. The processing heads for cutting, welding, and cladding are different and specific to the applications. For the consolidated control of the laser material processing, the use of integrated controller is the recent trend.

Though a large variety of lasers employing different kinds of active medium covering a wide range of wavelengths and powers has been developed, only few are being used for material processing. CO_2, Nd:YAG, and fiber lasers are the most popular systems, while excimer and diode lasers are also being employed for various other applications [2, 4]. Table 1.1 presents different type of lasers and their potential application.

In the field of industrial material processing, lasers have given a new direction to cutting of metallic and non-metallic sheets, welding of similar and dissimilar metals and composites, drilling, marking, metal forming, surface hardening, peening, surface alloying, cladding, and rapid manufacturing. Figure 1.6 presents the overview of laser-based manufacturing processes and their process domain in terms of laser power density and interaction time [2, 5]. Consequently, the scenario

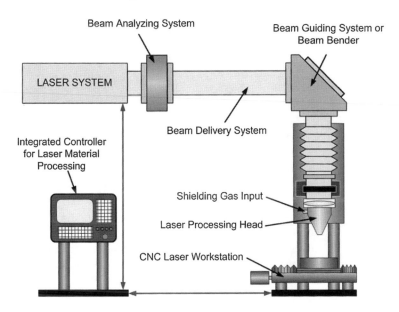

Fig. 1.5 Schematic arrangement of a typical laser processing station

Table 1.1 Different types of lasers and their applications

Laser (wavelength)	Applications
CO_2 laser (10.6 μm)	Light- to heavy-duty industrial cutting, welding, and rapid manufacturing
	Laser surface modification including cladding, alloying, and rapid manufacturing for wear/corrosion resistance and dimensional restoration
	Laser ablation, and laser glazing
Nd:YAG laser (1.06 μm)	Light- to heavy-duty job shops in drilling, welding, cutting, marking, and rapid manufacturing
Fiber laser (1.08 μm)	Laser cleaning in conservation of artifact, paint stripping
Excimer laser KrF (0.248 μm)	Optical stereolithography
XeCl (0.309 μm)	Marking, scribing, and precision micromachining involving drilling, cutting, etching of profile
Copper vapor laser (0.51 μm)	High-speed photography
	Detection of finger prints for forensic applications
	Excitation source for tunable dye laser for isotope separation
	Precision microhole drilling and cutting
Semiconductor laser/ diode laser	Optical computers, CD drivers, laser printers, scanners, and photocopiers
(0.8–1.0 μm)	Optical communication Industrial alignment
	Holography, spectroscopy, bio-detectors, ozone layer detector, pollution detection, bar code scanners, 3D image scanners

in manufacturing has changed, specifically for automobile, chemical, nuclear, and aerospace industries.

1.3.1 Laser Cutting

Laser cutting is one of the largest applications of lasers in metal-working industry and is a well-established universal cutting tool enables cutting of almost all known materials. One of the foremost reasons for wide acceptance of laser cutting was direct replacement of conventional cutting source with laser. When compared with other cutting processes (such as oxy-fuel cutting, plasma cutting, sawing and punching), its advantages are numerous, namely a narrow cut, minimal area subjected to heat, a proper cut profile, smooth and flat edges, minimal deformation of a workpiece, the possibility of applying high cutting speed, intricate profile manufacture and fast adaptation to changes in manufacturing programs [6]. The laser cutting uses different cutting mechanisms to cut different materials. Some of the mechanisms are vaporization, scribing, melt and blow, melt blow and burn, and thermal stress cracking [2].

Vaporization: In vaporization cutting, the focused beam heats the surface of the material to boiling point and generates a keyhole. The keyhole leads to a sudden increase in absorptivity and it results in quickly deepening the hole. As the hole deepens and the material boils, generated vapors erode the molten walls that ejects out the wall material, further enlarging the hole. Organic materials such as wood, Perspex, thermoset plastics, fiber-reinforced plastics are usually cut by this method.

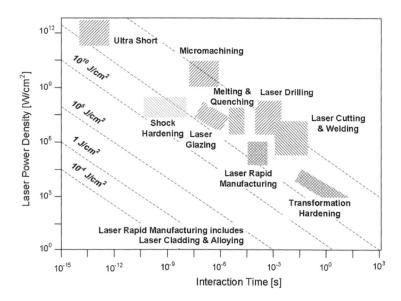

Fig. 1.6 Overview of various laser-based manufacturing processes and their process domain in terms of laser power density and interaction time [2, 5]

Melt and blow: Melt and blow or fusion cutting mode uses high-pressure gas to blow molten material from the cutting area, greatly decreasing the power requirement. First the material is heated to melting point then a gas jet blows the molten material out of the kerf avoiding the need to raise the temperature of the material any further. Materials cut with this process are usually metals. The distinct feature of this cutting mode is that the cut-edge material is same as the base metal and can be put to weld without cleaning the edge. The re-solidified layer has microcrack and ripple along the edge.

Scribing: This mode of cutting involves the drilling of small overlapping hole along the desired cut line by vaporization of the material. The sheet is then crack along the cut line. Generally, hard and brittle materials like ceramics and glass are cut by this mode of cutting.

Melt blow and burn: Melt blow and burn is a reactive cutting. This mechanism is successfully demonstrated in metals especially mild and stainless steel during oxygen-assisted cutting. This mechanism allows the cutting of very thick steel plates with relatively little laser power. The distinct feature of this cutting mode is that there is edge hardening due to existence of oxide and thermal cycle on the cut edge. This mode of cutting is also known as "burning stabilized laser gas cutting."

Thermal stress cracking: Brittle materials are particularly sensitive to thermal fracture, a feature exploited in thermal stress cracking. A beam is focused on the surface causing localized heating and thermal expansion. This results in a crack that can then be guided by moving the beam. The crack can be moved as fast as in order of few m/s. It is generally used in cutting of glasses.

The use of various mechanisms during laser cutting of various materials leads to various advantages over the conventional cutting. The cutting capacity for particular set of processing parameters can be estimated by severance energy (SE). SE is defined as follows:

$$SE = \frac{P}{Vt}$$

where P is laser power (W), V is cutting speed (mm/sec), and t is thickness of the material. The quality of laser cut, that is, width of laser cut or kerf and the quality of cut edges depend upon laser, laser power (average for continuous wave (cw) and peak for pulsed), the motion of laser beam and workpiece. Table 1.2 presents laser cutting kerf width versus material thickness for some of the important engineering materials [7].

With increasing demands of personal customization and to provide variations, the laser cutting with three-dimensional processing is being widely used in interior of cars and high-end buses (most popular application is in roof linings, door, instrument panels, and arm rests), cutting of complex pipe profiles and hydro-formed parts [7, 8]. Most of the parts used in above applications area are primarily made up of polymer-based materials, and hence, the availability of robots resulted in increased popularity of this technology.

Table 1.2 Laser cutting kerf width and material thickness for important engineering materials [7]

Material	Material thickness	Kerf width
Aluminum	2–3 mm	0.2–0.3 mm
Plastics	50–500 μm	2 × beam diameter
Steels	1.5 mm	50 μm
	2.5 mm	150 μm
	3.0 mm	200 μm
	6.0 mm	300 μm

Recent developments in laser systems in terms of power, beam quality, and different novel power modulation schemes have aided in addressing many complex problems of laser cutting in more productive way. Figure 1.7 presents the various profile cutting carried out at the authors' laboratory. The recent studies on the phenomenon of "striation free cutting," which is a feature of fiber laser assisted oxygen cutting of thin-section mild steel, concluded that the creation of very low roughness edges is related to an optimization of the cut front geometry when the cut front is inclined at angles close to the Brewster angle for the laser–material combination [9]. The studies to obtain fine and spatter-free pierced holes at the site of cut initiation along with dross free cut edges of minimum roughness and microstructural changes were also carried out. These studies indicated that suitable close-loop control of duty cycle in proportion to cutting speed in progressive change in pulse duty cycle in proportion to cutting speed will effectively suppress unwanted heating effects at sharp corners of laser cut profiles [10].

1.3.2 Laser Drilling

Since the invention of the laser, there has been a constant development to shorter pulse times. Not long ago 10-ns pulses were the shortest obtainable but now femtosecond lasers are widely applied and even shorter pulses can be obtained in

Fig. 1.7 Profile laser cutting carried out at RRCAT

the laboratory. When energy is released in very short time, it results in high peak powers as high as 10^{10} W or orders more. The intensity of the incoming beam is expressed as I_0. The decrease in the laser intensity into the depth of material is given by $I_x = I_0 e^{-\alpha x}$, where α optical absorptivity of the material and x the depth into the material. The optical penetration depth δ ($\delta = 2/\alpha$) is the depth of material whereby almost all laser energy is absorbed. This optical penetration depth for metals is found to be in the order of 10 nm. It means that the laser energy heats a 10-nm-thick layer of metal in 1 ps. This heat will diffuse from that skin layer (δ) to the bulk. The diffusion depth is expressed by $d = \sqrt{4 a t}$ with a as the thermal diffusivity and t the diffusion time. In case of steel, for 10-fs pulse, we obtain a diffusion depth of 1 nm while during a 1-ps pulse, the heat diffuses over 10 nm. Taking the results together than we see that

- It takes 1 ps to convert laser energy into heat.
- This takes place in a 10-nm-thick skin layer.
- The diffusion depth for 1 ps is also 10 nm.

From these results, we consider a pulse as ultrashort when the (thermal) diffusion depth during the pulse is in the same order or less than the skin layer depth (optical penetration depth). The optical penetration depth depends on the material and the laser wavelength. The diffusion depth depends on the material properties. Table 1.3 presents the ultrashort pulses common for some of the important engineering materials.

When very short pulsed laser beam is focused on any metallic surface, it simply ablates the thin layer of material due to very high power density. The single or repetitive use with appropriate position of focal spot of such pulse results in drilling. In terms of process technology, this technology can be deployed in four ways: single-pulse drilling, percussion drilling, trepanning, and helical drilling. Figure 1.8 presents the schematic diagram of these processes.

In single-pulse drilling, the hole is created in single pulse. Since the drilling is depending on single pulse, the pulse duration and beam profile plays a vital role. The attainable depths in single-pulse drilling range from several micrometers for nanosecond pulse and several millimeters for microsecond pulse. For drilling, higher depth repetitive pulses with appropriate position of focal spot are used and this process is known as percussion drilling. For drilling holes greater than the focal spot and of non-circular shape, the laser beam spot is translated in the profile plan and this drilling process is called laser trepanning. Helical drilling is special

Material	Ultrashort pulse
Metal	1 ps
Ceramic	10 ps
Plastic	1 ns

Table 1.3 Ultrashort pulse for some of the important engineering materials [7]

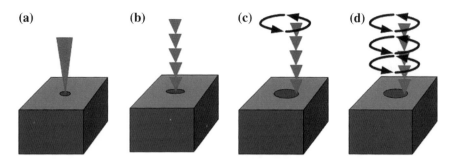

Fig. 1.8 Schematic diagram presenting different drilling processes **a** single drilling **b** percussion **c** trepanning, and **d** helical drilling

case of laser trepanning when the focal spot of laser beam is positioned appropriately along the drilling depth.

Drilling holes on aerospace and avionic components to improve the heat transfer, toil filers of automobiles, are some of the important industrial applications of laser drilling. As compared to electro-eroded and mechanically drilled holes, the laser-drilled holes still have limitations especially in terms of concentricity and burring. Laser drilling is also used in drilling holes in diamond, one of the hardest material on earth. There is difficulty in diamond drilling because it is transparent for a wide range of wavelength. At high power densities, however, the diamond is transformed into graphite, which absorbs the laser power and is removed by ablation subsequently. Diamond machining is currently done by microsecond pulses of Nd:YAG lasers and nanosecond pulses of excimer lasers [11]. Thin layers of graphite or amorphous carbon are found on the surface after laser machining which requires an extra polishing operation to remove the graphite. However, this extra step of polishing can be removed if ultrashort femtosecond lasers are used [12].

In the domain of microdrilling, Excimer lasers are preferred over other lasers, as they offer three significant advantages. First, the short ultraviolet light can be imaged to a smaller spot size than the longer wavelengths. This is because the minimum feature size is limited by diffraction, which depends linearly with the wavelength. The second advantage is that due to the mechanism of "photoablation" there is less thermal influence or melting of the surrounding material. Finally, most materials show a strong absorption in the ultraviolet region. This means that the penetration depth is small and each pulse removes only a thin layer of material, allowing precise control of the drilling depth [8, 13].

One of the exciting applications is fabrication of printed circuit board (PCB), where many bridging holes (vias) are produced to make electrical connections in multi-layer PCBs. The holes are drilled in dielectric polyimide layer until the underlying copper layer is uncovered. The drilling then stops automatically because of the higher threshold (one order of magnitude) of copper. The conducting connection is made by a following chemical deposition of copper on the walls. The process has been developed by Bachmann [14] and is used for drilling small ≈ 10 μm holes. For bigger holes of 100 μm and above, the cheaper and faster CO_2 lasers are currently used. In the fabrication of nozzle for the ink jet printers, an array of small orifice with precisely defined diameter and taper are required. These holes are located on the top of a channel with resister heater. Small bubbles are formed when the ink is heated, ejecting small (3–80 pl) drops out of the nozzle. Riccardi et al. [15] describes the fabrication of high-resolution bubble ink jet nozzles (similar application is referred in Fig. 1.9). Depending on the design, up to 300 holes have to be drilled simultaneously in a 0.5×15 mm area. The total drilling time is about 1 s, using a 300 Hz, KrF laser. Recent developments have facilitated the smaller holes below 25 μm diameter for this application.

1.3.3 Laser Welding

When high-power laser beam is focused, it produces very high intensities of the order of 10^4–10^7 W/cm^2 at the focal spot. When such a high-intensity spot is placed on the edges of the materials forming a joint in butt or lap joint configuration, the material melts and solidifies as soon as the beam is passed away. This melting and solidification of the edges under proper shielding result in welding of the material. Among various welding processes, high-energy beam welding employing laser has the key benefits in terms of localized heating, faster rate of cooling, smaller heat-affected zone (HAZ), easier access weld seam through fiber delivery, access to weld intricate geometrical shapes and sizes, reduced workpiece distortion and undulation and possibility of performing welding in ambient and controlled environment [16–21]. The following are main characteristics of the laser-welded joints:

Fig. 1.9 An injection nozzle hole with a high-power picosecond laser produces very sharp edges with no burr or melt and low surface roughness inside the hole, resulting in an optimum spray cloud of the fuel

20 μm

- Deep and narrow weld.
- Very low thermal distortion and residual welding stresses.
- Narrow HAZ and minimum metallurgical damage.
- High speed and high production rate.
- High precision control in space and energy.
- Autogenous welding, that is, no-filler is required.
- Possibility of dissimilar material joints.
- Welding of relatively remote- and limited-access locations.
- Easy to automate.

Because of these distinguished advantages, it is widely employed for welding coarse to fine precision of thick to thin metals, ceramics, and polymers. The applicability and preference of laser welding depends on laser properties and parameters, material properties and methodology, but the selection of welding joint is broadly governed by the functionality of components. There are few limitations of the laser welding:

- Close fitting of joints and accurate beam/joint alignment is required.
- Precision control of laser and process parameters are required.
- Fixed and running cost of the laser welding machines is high.

To form a laser weld, the laser beam is finely focused on the center line of the joint. Initially most of the incident laser power (even more than 90 %) is reflected, since at the room temperature, metals are good reflector of the infrared radiation. As the metal surface is heated, the surface reflectivity decreases until at the boiling, where it is negligible. The laser energy is absorbed in the skin depth and depending upon the power intensities, two different modes of the laser welding takes place. One is conduction welding and another is keyhole welding. When the material thickness is small and the intensity of the incident laser is relatively low, the material melts and temperature does not exceed the boiling temperature of the material. The aspect ratio (ratio of weld bead height to width) of the weld bead is between 1 and 1.5 for this mode of laser welding. When the power intensities are sufficiently high to cause the vaporization of the material, a hole is created at the center of the molten pool by rapidly escaping metal vapors. In this case, the laser beam further melts and vaporizes the material at the bottom of the created hole. Thus, a narrow hole is formed along the depth of the material. There is dynamic equilibrium among the molten metal, escaping vapor, heat input due to laser, and associated heat transfer phenomena. At the trails edge of the keyhole, the molten material collapses and solidifies forming a deep penetration welding. This mode of welding leads to high aspect ratio in the range of 6–10 (Fig. 1.10).

There has been an extensive experimental and modeling efforts giving insight into laser welding process and associated control for quality and repeatability. In laser welding, when metal melts, its volume increases due to thermal expansion and it forms a convex-shaped weld bead after solidification. The improper selection of

Fig. 1.10 Schematics of **a** heat conduction and **b** deep penetration welding

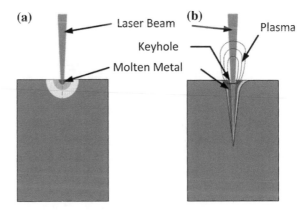

process parameters results in humping. In humping, there is periodic formation of convex and concave weld bead formation (Refer Fig. 1.11 for details) [22].

In many applications, this protruding convex-shaped weld bead is not acceptable. This can be reduced by appropriate edge preparation. In one of the recent efforts, the size and shape of the edge preparation were estimated experimentally in the authors' laboratory. 6-mm-thick sample of austenitic stainless steel-type 304 (size 15×100 mm) having chamfered edges with an included angle of $60°$ at various chamfered depth ($d = 0.5, 0.8$ and 1.0 mm) was used for the investigation. The detail of chamfered edges is presented in Fig. 1.12. During laser welding, the molten material has tendency to escape away from the molten pool due to its dynamics. Figure 1.13 presents the typical flow molten metal in the molten pool during laser welding. The selected chamfered edges provided a wider wetting surface and easy expansion of the molten material. The depth of the chamfered is selected to provide the compensation to convexity. Table 1.4 presents the results of the laser welding experiments. Fig. 1.14a and b presents the top view of weld sample without chamfered edges and with 0.8-mm-deep chamfered edges, respectively.

Fig. 1.11 Schematic diagram of humping

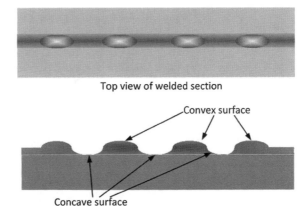

During the laser welding, there is angular distortion of the plates due to heating and cooling of the material near the weld zone. There have been several theoretical and experimental studies to predict the distortion [23, 24]. These studies are the extension of the similar studies carried out for the conventional welding processes. For butt-joint configuration, the distortion can be minimized by suitably selecting the proper joint geometry. The angular distortions of 6-mm austenitic stainless steel 304 plates were also studied, and an alternative methodology for minimizing distortion in butt-joint configuration was developed recently in authors' laboratory. Laser welding in butt-joint configuration for the sample with and without chamfered edges was carried out. It was found that the angular distortion for the samples with edges was significantly lesser than that of without chamfered edge. The angular deformation of laser-welded joints having different depth of chamfered edges as a function of laser energy per unit length is shown in Fig. 1.15. Minimum distortion was obtained with 0.8 mm groove depth with laser power per scan speed of 145 kJ/m. For further reduction in the angular distortion, the laser welding was tried from both top and bottom sides. Negligible angular distortion was obtained for both sides welding. The effect of groove on deformation is presented in Fig. 1.16a and b. It is just because the thermal stresses generated during the laser welding cancels each other when welded from the both sides.

Recently, a new approach of laser welding was reported by Fraunhofer IWS Dresden to weld dissimilar materials such as aluminum/copper, aluminum/magnesium, or stainless steel/copper clearly showing better quality [25]. They deployed a highly dynamic 2D scanner with high scanning frequency (up to 2.5 kHz) to generate extremely small weld seam with high aspect ratio. This leads to very short melt pool lifetime, thereby suppressing the formation of brittle intermetallic phases. In these experiments, the melting behavior of metallic mixed joint, seam geometry, chemical composition, melt pool turbulence, and solidification was controlled by high frequency, time, position, and power-controlled laser beam oscillation. Using this strategy, phase seam values less than 10 μm was obtained for the aluminum/ copper dissimilar joint. The tensile strength of this dissimilar joint was found to be same value as that of aluminum/aluminum joint [25].

1.3.4 Laser Brazing

Dissimilar metals are preferred due to better material utilization with improved functionality in many engineering applications. It has encouraged the research thrust on various brazing processes including laser brazing. Among the various

Fig. 1.12 Details of chamfered edges preparation (d = depth of groove)

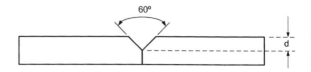

Fig. 1.13 Typical Marangoni flow of molten metal due to chamfered edge during laser welding

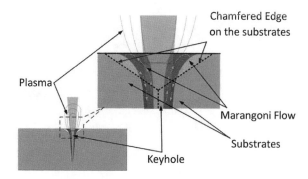

Table 1.4 Observed weld bead profile at various processing parameters during laser welding at authors' laboratory

S. No	Laser power (kW)	Scan speed (m/min)	Groove depth (mm)	Observed profile of weld bead
1	2	0.7	Without groove	Convex
2	2	0.8	Without groove	Convex
3	2	0.9	Without groove	Convex
4	2	1.0	Without groove	Convex
5	2	1.1	Without groove	Convex
6	2	0.7	1 mm	Small convex
7	2	0.8	1 mm	Flat
8	2	0.9	1 mm	Concave
9	2	1.0	1 mm	Concave
10	2	1.1	1 mm	Concave
11	2	0.5	0.8 mm	Concave
12	2	0.6	0.8 mm	Concave
13	2	0.7	0.8 mm	Flat
14	2	0.8	0.8 mm	Small convex
15	2	0.9	0.8 mm	Convex
16	2	1.0	0.8 mm	Convex
17	2	1.1	0.8 mm	Convex
18	2	0.7	0.5 mm	Convex
19	2	0.8	0.5 mm	Convex
20	2	0.9	0.5 mm	Convex
21	2	1.0	0.5 mm	Small concave
22	2	1.1	0.5 mm	Concave

combinations of materials, Cu-SS combination is important as Cu has higher thermal conductivity while SS has higher strength. This joint finds application in many advanced engineering applications, like particle accelerators and power plants where efficient removal of heat is mandatory along with material strength. Joining of AISI-type 304L stainless steel with copper using conventional processing portrays multi-fold problems due to the difference in thermophysical properties, while it is challenging for laser processing due to high reflectivity of Cu

(a)   (b)

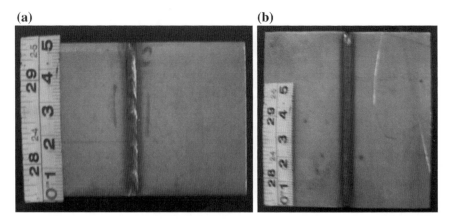

Fig. 1.14 a Weld bead without groove and undulation. **b** Weld bead With groove and controlled undulation

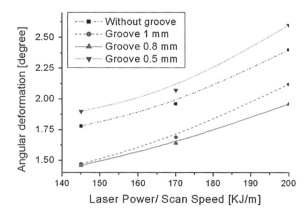

Fig. 1.15 Angular deformations at different laser energy level per unit length

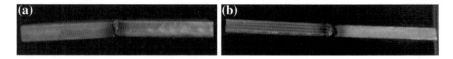

Fig. 1.16 a Side view of welding sample without chamfered edge exhibiting the angular distortion. **b** Side view of welding sample with chamfered edge exhibiting reduced angular distortion

[26, 27]. Various brazing techniques are used to join two dissimilar materials [28–30]. In these processes, the closely fitted materials/parts with filler metal are heated above 450 °C, facilitating molten filler metal to flow into fine gaps by capillary action to form material joint. Conventionally, the brazing is performed in a

controlled atmosphere furnace or in vacuum chamber [31] for the joining of Cu with SS. The furnace brazing is very slow process with the unnecessary heating of whole job at high temperature. This heating may produce high thermal stresses in the parts and can change the surface properties of base materials. Laser is widely used as heat source in many material processing applications, including cutting, welding, cladding, surface treatment, and rapid manufacturing [2, 32, 33]. When this is used for the brazing process, it is called as laser brazing [34–36]. By using laser brazing, very thin and miniature parts can be joined with well-controlled HAZ. In the present study, a 2-kW ytterbium fiber laser integrated with 5-axis workstation was used for laser brazing. A number of samples were brazed with using active brazing filler foil (Ag-63 Cu-35.25 Ti-1.75) of thickness 50 microns in butt-joint configuration. The brazed joints were subjected to various non-destructive (visual and dye-penetrant test) and destructive (microscopic examination, energy-dispersive spectroscopy and four-point bending test) characterization techniques.

Two basic process parameters, that is, laser power (P) and scan speed (v), were varied, and their effect on the brazed joint was studied at authors' laboratory. At laser power below 350 W and moderate scan speed (4 mm/min), the proper melting of brazing foil was not occurred. As we increased the laser power keeping the scan speed constant, the melting of filler foil observed and the brazing of Cu-SS was witnessed with relatively good strength. Experiments were also carried out to evaluate the effect of scan speed at constant laser power (450 W) on laser brazing. It was observed that when scan speed was 6 mm/min, the proper melting of filler foil was not occurred. As the scan speed decreased from 6 mm/min, the melting of foil was observed but still the joint was not good by appearance. As scan speed is reduced to 2 mm/min, the melting of foil was occurred and a better brazed joint was observed. The optimum processing parameters for laser brazing of 3-mm SS-Cu joint was found to be 450 W and 2 mm/min.

A number of non-destructive and destructive tests were carried out to characterize the brazed joints. Visual examination is the preliminary examination carried out after laser brazing to detect surface defects, distortion, bead appearance, lack of penetration, spattering etc. The visual examination for the various brazed joints was carried out and the defect-free joints were taken for further examinations. Dye-penetrate test is also carried out for the selected samples and found them with no leaks. This constitutes one of the most important tests, capable of providing insight into modifications taking place in the material as results of laser processing. The feedbacks received from microstructural analysis are often used for optimizing laser processing parameters. The test can provide wide ranging information, for example, microstructural changes, development of deleterious phases, if any (especially during dissimilar metal welding), extent of laser-affected zone, nature and extent of defects developed in the laser-processed joint. Figure 1.17 presents microscopic examination of typical laser-brazed samples at two laser powers at constant scan speed (2 mm/min).

For microstructural examination, brazed samples were polished by various metallurgical techniques and prepared for energy-dispersive X-ray spectroscopy.

Elemental distribution across the brazed joint was studied through point and line scanning. Cu substrate that is used for the brazing experiments had 99.9 % purity. Figure 1.18 presents the elemental distribution across the laser-brazed joint of SS-Cu. It is evidently clear that Fe signal falls as we move from the SS to Cu. There is rise in Cu signal and a peak of Ag in the middle of SS and Cu. The gradient at the interface is sharp, it reconfirm that there was no dilution of the SS and Cu substrate during the brazing process. The line scan of Ag is displaced toward the Cu, as the wettability of Ag is more on Cu as compared to SS.

To evaluate the joint strength, four-point bending test of the laser-brazed joints were conducted and flexural joint strength was evaluated. Prior to testing of brazed joints, four-point bending test specimen was machined from as-received material for evaluating the flexure strength of the material in as-received condition. Table 1.5 summarizes the results of four-point bending test of material in as-received condition. As the laser energy per unit length is increased, the flexural strength of brazed joint increases. It is because the availability of more energy per unit length will result in proper melting and wetting of the brazed material across the joint. As the thermal conductivity of Copper is more and melting point is lower as compared to corresponding values of Stainless steel. The effect of laser focal spot offset in transverse plane is also investigated. The laser beam was offset by 0.5 mm toward SS and Cu and brazing was carried out. When we offset the laser beam toward the Cu side by 0.5 mm, the flexural strength decreases for the laser-brazed joint at various combination of laser power and scan speed. This is due to the availability of lower laser energy (because of high reflectivity at Cu surface and higher thermal conductivity of Cu material) yielding to insufficient wetting of to-be-brazed surfaces. When the laser beam was offset toward the SS side by 0.5 mm, the flexural strength of brazed joint was decreased as compared to that of with zero offset but the flexural strength was increased as compared to that of with laser beam offset toward the Cu side. This is due to the less reflection of laser beam and less thermal conductivity of SS as compared to Cu yielding to relativity higher wetting. Further, the flexural strength is low as compared to zero offset beam

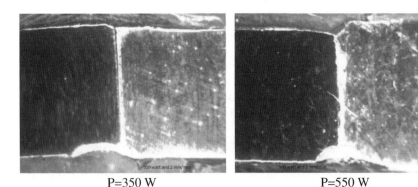

P=350 W   P=550 W

Fig. 1.17 Pictograph of transverse cross section of laser-brazed samples of 3-mm-thick sheets for constant scan speed of 2 mm/min at various laser powers

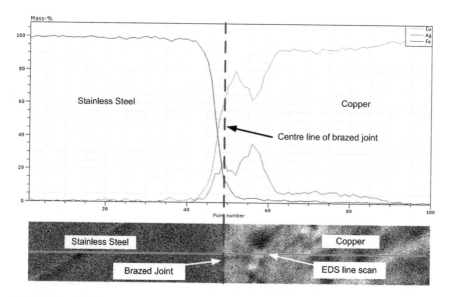

Fig. 1.18 Typical elemental distribution across the laser-brazed joint of SS-Cu

brazing joint because offset of laser beam effects the melting of brazing foil and the heating of contacted surfaces. Results also indicate that the offset of laser beam is very sensitive as it affects the strength of joint. So the alignment and laser beam positioning should be done very carefully (Figs. 1.19, 1.20).

Laser brazing is an attractive technique of joining the material when heat input and subsequent distortion is one of the major criteria for the fabrication of engineering component. In our present study, the laser brazing of Cu and SS was done and the results were compiled. The visual inspection and dye-penetrate testing were done and found that the joint was good. The optical microscopy for the various joints was done. Further, the flexural testing using four-point bending using UTM was carried out and the flexural stresses for different brazed joints were determined and compared with base materials. The scanning electron microscopy was carried out for the various joints and an analysis of the element distribution was performed for material characterization. Maximum flexural stress observed was 343 MPa at the laser power of 550 W and the scan speed of 3 mm/min. If we further increase the laser power, the samples were observed with more

Table 1.5 Results of four-point bending test of material in as-received condition

Sample	Flexural strength [MPa]	Average flexural strength [MPa]
Copper	579.07	506.82
Copper	434.57	
SS 316L	838.83	748.97
SS 316L	659.11	
Brazing foil		450[a]

[a] As per manufacturer's material test report

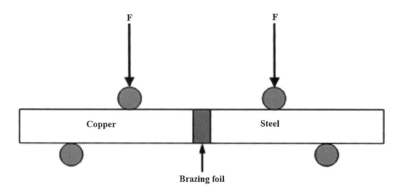

Fig. 1.19 Schematics depicting four-point bending test setup

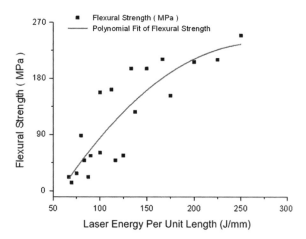

Fig. 1.20 Effect of laser power and scan speed on flexure strength of brazed joint of 3-mm SS-Cu sheets

HAZ and some melting of base metal copper was observed at 600-W power with 2- and 3-mm/min scan speeds. At low laser power (less than 350 W), the brazing foil was not melted fully even the scan speed was as low as 2 mm/min. The scanning electron microscopic analysis shows that the diffusion of Ag component of brazing foil is greater in the copper side than SS side.

1.3.5 Laser Rapid Manufacturing

LRM is one of the advanced additive manufacturing processes, that is, capable of fabricating engineering components directly from a solid model. This process utilizes the laser energy as a heat source to melt a thin layer of the substrate together with the blown metal powder to form a layer in a predetermined shape. A number of such layers deposited one over another, and subsequently, it results in fabrication of three-dimensional (3D) components directly from the solid model.

LRM eliminates many production steps such as assembly, man–machine interaction process planning, intermittent quality checks, and consequently related human errors. It also offers many advantages due to its inherent manufacturing features over the conventional subtractive techniques such as reduction in production time and fabrication of functionally graded parts (complex heterogeneous/porous structures). LRM is now put to commercial use for the fabrication of few functional metal parts. Conventional operations show obvious advantages for fabrication of parts in large volume. But when the part is unusual in shape or has fine internal features, the turnaround and cost will increase rapidly, and in some cases, it is impossible to realize. Price is almost secondary to function in some cases. Aerospace, military, space programs, certain marine applications, and biomedical implants come under this category. For these high-value prime applications, LRM is most appropriate. Because of the process flexibility (in terms of raw material selection and feeding), it is possible to make components with compositional grading across length/width to meet the multi-functional requirements. Thus, fabrication of functionally graded materials is one of the important capabilities of rapid manufacturing which finds applications in both the engineering and prosthetic components.

Features that can be attributed to a machine part include the geometry of the part, the material composition, and the microstructure. The material composition is governed by the efficiency of the material delivery method, whereas the microstructure is governed by the specific energy input. As laser-based additive manufacturing techniques can address all these complex issues, manufacturing techniques similar to LRM are being developed with different names at various laboratories around the world. At Sandia National Laboratory, USA, laser engineered net shaping (LENS^{TM}) is being developed with prime focus on creating complex metal parts in single day [37]. National Research Council, Canada, is developing freeform laser consolidation for manufacturing of structural components for advanced robotic mechatronic systems [38]. Automated laser fabrication (ALFa) is being developed to produce low-cost tungsten carbide components at the University of Waterloo, Canada [39]. Selective laser cladding (SLC) at the University of Liverpool, UK, and direct metal deposition at the University of Michigan, USA, are being used for depositing critical surfaces on prime components [40, 41]. Laser powder deposition (LPD) at the University of Manchester, UK, and direct metal deposition/laser additive manufacturing at Fraunhofer Institute, Germany, are being augmented for the fabrication of high-performance materials [42]. The researchers at Tsinghua University, China, are working on diverse area and evaluating the potential of technology for the development of graded Ti alloys for aeronautical, nickel alloys for power plants and various in-situ repair applications [43]. Thus, the ongoing global research is spearheading toward the deployment of the fabrication technology for improving qualities of the products integrated with multi-materials and multi-functional components enhancing a step benefit in economics.

Lasers in LRM: In LRM, high-power laser system is used as heat source to melt thin layer of substrate/previously deposited layer and fed material. CO_2, Nd:YAG, and diode lasers are most widely used for the application. The availability of high-power fiber lasers made it a new entrant in this application domain [44]. Since the wavelength of Nd:YAG, diode, and fiber lasers are near 1 µm, the absorption is better for laser material processing involving metals as compared to that of CO_2 lasers. However, CO_2 lasers are still being widely used due to less cumbersome safety infrastructure, established systems and procedures. As against the common notion, both pulsed and cw lasers have been successfully used for LRM. The laser energy intensity of 20–60 kW/cm^2 is used for CO_2 lasers, while it is 150–200 kW/cm^2 for pulsed Nd:YAG lasers [45, 46]. The basic prerequisite for laser beam energy intensity distribution is symmetry along the axis of laser beam propagation. It allows uniform material deposition independent of direction of processing. Therefore, multimode laser beam with flat top distribution is most widely used. Gedda et al. compared the laser absorption and the energy redistribution during LRM process using CO_2 laser and Nd:YAG laser [47].

System requirements: LRM system consists of the following three primary subsystems:

1. High-power laser system: One of the following lasers is commonly integrated within the LRM system: CO_2 laser; Nd:YAG laser; diode laser; fiber laser
2. Material feeding system: Among material feeders, there are three main types of feeding techniques. They are wire feeding, preplaced powder bed and dynamic powder blowing.
3. Computerized Numerically Controlled (CNC) workstation: It is either 3- or 5-axes workstation.

The details about the lasers used in LRM systems are already presented in Fig. 1.21. Among the material feeding systems, wire feeders are directly adopted from metal inert gas welding (MIG) process. Wire feeding is preferred for the fabrication of components involve continuous deposition [48, 49], as intermittent start/stops results in discontinuity in deposited material. This method is not adopted as universal method, because of poor wire/laser coupling leading to poor energy efficiency, unavailability of various materials in wire forms and their cost. In preplaced powder bed, a predefined thickness of the powder is laid on the substrate and the powder is melted using laser to form the solidified layer. The method is preferred to fabricate fine features and overhang structures. The method has limitation in achieving cent percent density in the deposits. Dynamic powder blowing is convenient and most widely used approach for material feeding in LRM systems. It allows online variation in feed rate and multi-material feeding. Moreover, laser energy utilization is also more in dynamic powder blowing, as the laser beam passes through the powder cloud to the substrate/previously deposited layer, resulting in preheating of powder particles by multiple reflections. Since the powder is fed into the molten pool for melting and forming layer, some powder particles are ricocheted from the pool. The ratio of the powder deposited to powder fed is termed as powder catchment efficiency. The typical achieved powder catchment efficiency is 35–80 %. In CNC workstations, 3-axis interpolation (X, Y and Z) is sufficient to reach to any point in the space, but two more axes (*A* and *C*) are required to orient particular direction to reach. Therefore, 3-axis configuration is minimum system requirement and 5-axis is universal requirement without redundancy. Apart from axes-movement, laser workstation needs some more features, like laser on/off, gas on/off, powder feeder on/off, for LRM.

Figure 1.22 presents LRM facility depicting an integrated LRM station with scope for glove box operation to fabricate components in a controlled atmosphere developed at authors' laboratory. Research is underway to incorporate the feedback-based, closed-loop control to precisely manage the temperature, cooling rate, size of the melt pool, and shape of the parts being manufactured. At authors' laboratory, a comprehensive program was undertaken to address various industrial and in-house applications. The fabrication of Colmonoy-6 bushes, solid and porous structures of Inconel-625, and low-cost WC–Co tools are the important applications among them [50–54]. The properties and performance of laser rapid manufactured components/structures are found to be at par with that of conventionally processing component.

Development of high performance surface on SS316L for improved wear resistance: Recently, the studies were extended to the development of high-performance surfaces by depositing tungsten carbide (WC)-reinforced nickel matrix on SS316L. In the study, Inconel-625 alloy (particle size: 45–105 μm) was used to provide nickel matrix for reinforcing WC particles (particle size: 45–75 μm). A number of test trails were made to optimize the processing parameters for LRM of continuous multi-layer overlapped deposition at various powder compositions. It was observed that the deposit got delaminated during the LRM of second layer for 5 % weight Inconel-625 due to insufficient wetting material. This problem was not

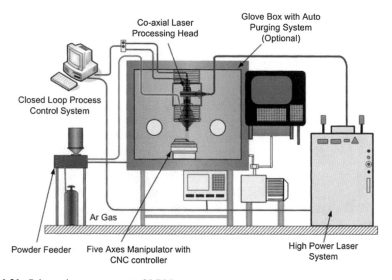

Fig. 1.21 Schematic arrangement of LRM setup

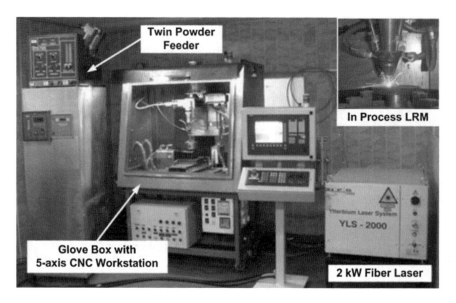

Fig. 1.22 Laser rapid manufacturing facility at author's laboratory

observed for higher compositions. Figure 1.23 presents a typical laser rapid manufactured WC-reinforced nickel matrix on 316L Stainless Steel.

The microscopic examination revealed that the dendritic microstructure of Ni-matrix originated from WC particles due to directional quenching (refer Fig. 1.24). The erosion wear performance of the laser rapid manufactured samples was evaluated using Al_2O_3 air-jet erosion test rig for Inconel-625 in deposit ranging 5–25 % weight, erodent jet velocity ranging 10–50 m/s, jet impinging angle ranging 0–68°, and substrate temperature 50–250 °C. Figure 1.25 presents contours depicting erosion wear rate (EWR) for various composition and impinging angle at erodent velocity 30 m/s and substrate temperature 150 °C. The study demonstrated that WC-reinforced Ni-matrix laser rapid manufactured with

Fig. 1.23 Typical laser rapid manufactured WC-reinforced nickel matrix on 316L SS

18 wt % of Inconel-625 has least EWR for the range of parameters under investigation and it was found to be nine times lower than that of bare SS316L surface.

In another study, high-performance layers of cobalt-free materials were developed and their cavitation and slurry erosion behaviors were studied. The preference to the cobalt-free materials for nuclear applications to avoid induced radioactivity was the main motivation to develop laser cladding of Colmonoy-5 (a nickel base alloy) and Metco-41C (an iron base alloy) on AISI-type 316L stainless steel substrate. The process parameters were optimized for developing continuous and defect-free laser cladding of Colmonoy-5 and Metco-41C on AISI-type 316L stainless steel substrate using dynamic powder blowing methodology (powder particle size used: 45–105 μm). The observed optimum parameters were as follows: laser power: 1.6 kW; scan speed: 0.6 m/min; and powder feed rate: 8 g/min, with 60 % overlap index.

Figure 1.26 presents the typical microstructure at substrate-clad interface for Colmonoy-5 and Metco-41C. X-ray diffraction studies showed the presence of carbides, borides, and silicides of Cr and Ni for Colmonoy-5 whereas Metco-41C cladding exhibited the work hardening microconstituent, that is, austenite. The microstructural studies showed that the clad layers of Colmonoy-5, Metco-41C, and Stellite-6 primarily contain very fine columnar dendritic structure while clad-substrate interface exhibited planar and non-epitaxial mode of solidification due to high cooling rates. Multi-pass cladding showed refined zone at interface due to remelting and solidification. Figure 1.27 presents the microhardness profile for these claddings.

The improvement in cavitation and slurry erosion resistance of laser cladding with respect to AISI-type 316L is summarized in Table 1.6. Metco-41C showed better cavitation and slurry erosion resistance (especially @ 30°) than other cladding used in this work. The improvement in erosion resistance of developed clad layers was primarily attributed to arrest of craters formed by the metal matrix due to high toughness and reasonably good hardness.

Fig. 1.24 Micrograph depicting the WC particles in nickel matrix with dendrites

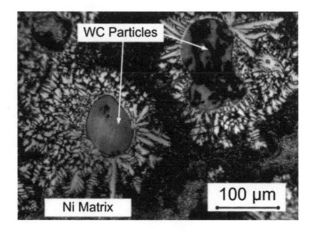

Fig. 1.25 Contours depicting EWR for various composition and impinging angle at erodent velocity 30 m/s and substrate temperature 150 °C

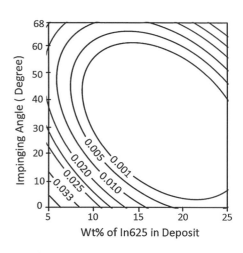

Wt% of In625 in Deposit

1.4 Laser Hazards and Safety

The lasers have been deployed widely for various manufacturing processes, but there are many laser hazards, and proper arrangements are prerequisite at the shop floor to avoid accidental damage, specifically to human [55]. Excessive levels of laser radiation are hazardous to any exposed area of human body. Because of their susceptibility to damage, the hazards to eye, skin, and the internal organs are different.

The eye hazards deserve special attention because of the particular vulnerability of eye and the extreme importance of sight. The general structure of the human eye is presented Fig. 1.28.

Light passes through the various ocular structures (*C*, *P* and *L*) to fall on the retina (*R*) where it triggers a photochemical process which evokes the neural impulses that leads to the perception of vision. Damage to the eye can occur on any of these structures and depends upon which structure absorbs the greatest

(a) **(b)**

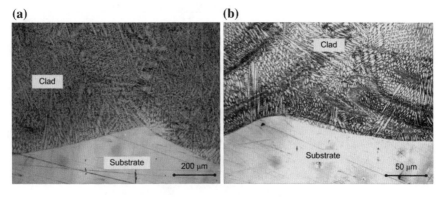

Fig. 1.26 Typical microstructure at clad-substrate interface for **a** Colmonoy-5 and **b** Metco-41C

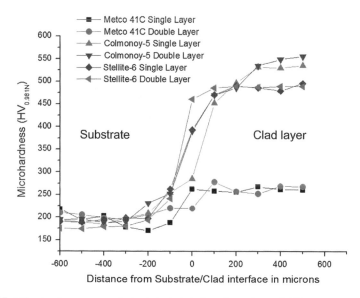

Fig. 1.27 Microhardness across clad-substrate interface for various cladding

Table 1.6 Wear behavior with respect to bare AISI-type 316L stainless steel

Description	Colmonoy-5	Stellite-6	Metco-41C
Cavitation erosion	1.6	4.1	4.6
Slurry erosion @ 30°	1.5	1.75	4.9
Slurry erosion @ 90°	2.2	3.5	2.9

radiant energy per unit volume of the tissue. The ocular media are transparent to radiation in the range 400–1400 nm. All the radiations falling on the cornea in this spectral range will be brought to a focus on the retina resulting very high retinal irradiance and it leads to retinal damage. The laser spot diameter on the retina may be very small (~ 1–10 µm), and it results in retinal irradiance up to 10,000 times higher than the corneal irradiance. The typical result of a retinal injury is blind spot within the irradiated area. The cornea is susceptible to damage by ultraviolet radiation (wavelength less than 300 nm) and from infrared radiation. If the injury is restricted to the outer most layers (~ 50 µm thick), the injury is likely to be very painful and gets heal within 48 h, it is because the outermost layer of the cornea gets renewed in period of every 48 h. However, the damage to the deeper layer of the cornea results in permanent corneal opacities and thus is very serious hazard.

Hence, the use of appropriate safety goggles in the laser laboratory is primary requisite before any laser experiments [56].

The skin injury thresholds are comparable with that of eye injury thresholds except the retinal injury where the radiation is focused on retina. However, the skin injury is normally very superficial unless incident takes place at very

Fig. 1.28 The general
structure of human eye

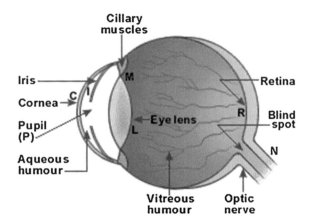

Fig. 1.28 The general structure of human eye

high-power level. It normally involves changes to the outer dead layer of the skin cell. The greater the penetration depth means the greater the volume of tissues available to deposit the absorbed energy. Because of this reason, the damage thresholds for visible and near infrared radiation are higher. Ultraviolet radiation induces photochemical reactions leading to erythema (reddening of skin) and prolonged exposure may initiate longer term degenerative processes, like accelerated skin aging and increased risk of certain type of skin cancer. Hence, appropriate guards and protective screen must be used for laser experiments to avoid direct or indirect exposure to skin.

Hazard to *deep lying organs* are rather unlikely and not generally considered harmful. It is suggested that the thicker hair, scalp, and skull of man would protect the human brain against the focused or unfocused laser beam in the range of 40 J.

Considering the health hazard with the exposure of laser radiation, the following general practices must be used in the laboratory or industry to avoid unwanted incidents [56]:

- Everyone who uses a laser should be aware of the risks. This awareness is not just a matter of time spent with lasers; to the contrary, long-term exposure with invisible risks (such as from infrared laser beams) tends to reduce risk awareness, rather than to sharpen it.
- Optical experiments should be carried out on an optical table with all laser beams traveling in the horizontal plane only, and all beams should be stopped at the edges of the table. Users should never put their eyes at the level of the horizontal plane where the beams are in case of reflected beams that leave the table.
- Watches and other jewelry that might enter the optical plane should not be allowed in the laboratory. All non-optical objects that are close to the optical plane should have a matte finish in order to prevent specular reflections.
- Adequate eye protection should always be required for everyone in the room if there is a significant risk for eye injury.

- High-intensity beams that can cause fire or skin damage (mainly from class 4 and ultraviolet lasers) and that are not frequently modified should be guided through tubes.
- Alignment of beams and optical components should be performed at a reduced beam power whenever possible.

1.5 Conclusion

In manufacturing, the lasers have demonstrated the processing capability with improved quality in all three processing domains: material removal (cutting and drilling), material joining (welding and brazing), and material addition (cladding, alloying, and rapid manufacturing). In material removal, the laser cutting and laser drilling processes could do what was not possible with conventional processing. Deployment of lasers for nuclear decontamination and decommissioning is one of such practical examples. The laser brought the joining process to new height by welding dissimilar materials without sacrificing the relevant mechanical properties. The use of laser brazing also brought the little distortion without compromising the esthetic aspects of the esteemed products, like high-value cars. LRM is an extremely flexible technique with application in multiple areas from repair of large-scale components to manufacturing of component with specific end application. Availability of compact high-power lasers, advanced CAD/CAM systems with faster computing speeds, and advanced diagnostic and control systems has provided a new dimension to manufacturing, and LRM is one of such development. Technical factors, such as advancement in sub-systems, and economic factors, such as falling price of lasers and other sub-systems, will further alleviate the deployment of LRM technology to manufacturing.

With the increased interests from various industries, laser-based manufacturing is likely to lead all fields including aerospace, medical devices, and tooling. In combination with innovative design and planning, the capabilities of laser-based manufacturing have been established to fabricate complex components with delicate details that are very difficult or even impossible to make using conventional manufacturing processes. It does not mean that laser-based manufacturing is a threat to the existence of conventional manufacturing processes, but it is simply going to augment the industries with advanced manufacturing technology to address unresolved complex geometrical and material issues. In combination with other conventional manufacturing processes, laser-based manufacturing is going to provide a unique cost-effective solution to next generation "feature-based design and manufacturing" using the strength of virtual and remote manufacturing. In these ways, these technologies will augment each other to bring the new era of hybrid technologies.

Acknowledgments The authors express their sincere gratitude to Dr. P. D. Gupta, Director Raja Ramanna Centre for Advanced Technology (RRCAT) for his constant support and encouragement. Thanks are due to our collaborators Prof. A. K. Nath of Indian Institute of Technology, Kharagpur, India, and Prof B. K. Gandhi of Indian Institute of Technology, Roorkee, India. During the experimental work presented above, the technical support of Mr. Sohanlal, Mr. A. S. Padiyar, Mr. S. K. Mishra, Mr. C. H. Prem Singh, Mr. M.O. Ittoop, Mr. Abrat Varma, Mr. Anil Adbol, Mr. Ram Nihal Ram, Mr. P Sangale, and Mr. S. K. Perkar of RRCAT and Mr. N. Yadaiah of Indian Institute of Technology, Guwahati, is thankfully acknowledged.

References

1. Bertolotti M (2005) The history of the laser. The Institute of Physics, London
2. Steen WM, Mazumder J (2010) Laser material processing. Springer, London
3. Silfvast WT (2004) Laser fundamentals. Cambridge Press, Cambridge
4. Thyagrajan K, Ghatak A (2010) Lasers: fundamentals and applications. Springer, London
5. Meijer J (2004) Laser beam machining (LBM), state of the art and new opportunities. J Mater Process Technol 149:2–17
6. Dahotre NB, Harimkar SP (2010) Laser fabrication and machining of materials. Springer, London
7. www.obrusn.torun.pl/htm0/prod_images/rofin/Laserbook.pdf. Accessed 9 Dec 2012
8. Ready JF (2001) LIA handbook of laser material processing. Magnolia Publishing, Inc., Magnolia
9. Powell J, Al-Mashikhi SO, Kaplan AFH, Voiseya KT (2011) Fibre laser cutting of thin section mild steel: an explanation of the 'striation free' effect. Opt Lasers Eng 49:1069–1075
10. Kukreja LM, Kaul R, Paul CP, Ganesh P, Rao BT (2012) Emerging laser materials processing techniques for future industrial applications. In: Majumdar JD, Manna I (eds) Laser-assisted fabrication of materials. Springer, London
11. Windholz R, Molian P (1997) Nanosecond pulsed excimer laser machining of CVD diamond and HOPG graphite. J Mater Sci 32:4295–4301
12. Shirk MD, Molian PA (1998) Ultrashort laser ablation of diamond. J Laser Appl 10:64–70
13. Liu L, Chang CY, Wu W, Pearton SJ, Ren F (2013) Circular and rectangular via holes formed in SiC via using ArF based UV excimer laser. Appl Surf Sci 257:2303–2307
14. Bachmann FG (1990) Industrial laser applications. Appl Surf Sci 46:254–263
15. Riccardi G, Cantello M, Mariotti F, Giacosa P (1998) Micromachining with excimer laser. CIRP Ann 47:145–148
16. Quintino L, Costa A, Miranda R, Yapp D, Kumar V, Kong CJ (2007) Welding with high power fiber lasers—a preliminary study. Mater Des 28:1231–1237
17. Kinoshita K, Mizutani M, Kawahito Y, Katayama S (2006) Phenomena of welding with high-power fiber laser. Paper #902 proceedings of ICALEO, pp 535-541
18. Khan MMA, Romoli L, Fiaschi M, Sarri F, Dini G (2010) Experimental investigation on laser beam welding of martensitic stainless steels in a constrained overlap joint configuration. J Mater Process Technol 210:1340–1353
19. Katayama S, Kawahito Y, Mizutani M (2010) Elucidation of laser welding phenomena and factors affecting weld penetration and welding defects. Physics procedia 5:9–17
20. Zhang X, Ashida E, Katayama S, Mizutani M (2009) Deep penetration welding of thick section steels with 10 kW fiber laser. Transactions of JWRI 27:63–73
21. Manonmani K, Murugan KN, Buvanasekaran G (2007) Effects of process parameters on the bead geometry of laser beam butt welded stainless steel sheets. Int J Adv Manuf Technol 32:1125–1133
22. Berger P, Hügel H, Hess A, Weber R, Graf T (2011) Understanding of humping based on conservation of volume flow. Physics Procedia 12:232–240

23. Wang R, Rasheed S, Serizawa H, Murarkawa H, Zhang J (2008) Numerical and experimental investigations on welding deformation. Trans JWRI 37:79–90
24. Deng D, Liang W, Murakawa H (2007) Determination of welding deformation in fillet-welded joint by means of numerical simulation and comparison with experimental measurement. J Mater Process Technol 183:219–225
25. Kraetzsch M (2012) Laser beam welding with high frequency. LIA Today 18
26. Torkamany MJ, Tahamtan S, Sabbaghzadeh J (2010) Dissimilar welding of carbon steel to 5754 aluminum alloy by Nd:YAG pulsed laser. Mater Des 31:458–465
27. Anawa EM, Olabi AG (2008) Optimization of tensile strength of ferritic/austenitic laser welded components. Opt Laser Technol 46:571–577
28. Yoshihisa S, Takuya T, Kazuhiro N (2010) Dissimilar laser brazing of boron nitride and tungsten carbide. Mater Des 31:2071–2077
29. Mai TA, Spowage AC (2004) Characterization of dissimilar joints in laser welding of steel–kovar, copper–steel and copper–aluminium. Mater Sci Eng, A 374:224–233
30. Dharmendra C, Rao KP, Wilden J, Reich S (2011) Study on laser welding–brazing of zinc coated steel to aluminum alloy with a Zinc based filler. Mater Sci Eng, A 528:1497–1503
31. Lippmann W, Knorr J, Wolf R, Rasper R, Exner H, Reinecke A-M, Nieher M, Schreiber R (2004) Laser joining of silicon carbide—a new technology for ultra-high temperature resistant joints. Nucl Eng Des 231:151–161
32. Paul CP, Bhargava P, Kumar A, Kukreja LM (2012) Laser rapid manufacturing: technology, applications, modeling and future prospects. In: Davim JP (ed) Lasers in manufacturing. Wiley-ISTE, London
33. Masato K (2009) Fiber lasers: research, technology and applications. Nova Science, New York
34. Huang S, Tsai H, Lin S (2004) Effects of brazing route and brazing alloy on the interfacial structure between diamond and bonding matrix. Mater Chem Phys 84:251–258
35. Peyre P, Sierra G, Deschaux-Beaume F, Stuart D, Fras G (2007) Generation of aluminium–steel joints with laser-induced reactive wetting. Mater Sci Eng, A 444:327–338
36. Li L, Feng X, Chen Y (2008) Influence of laser energy input mode on joint interface characteristics in laser brazing with Cu-base filler metal. Trans Nonferrous Met Soc China 18:1065–1070
37. http://www.sandia.gov/mst/pdf/LENS.pdf. Accessed on 10 Dec 2010
38. Xue L, Islam MU, Theriault A (2001) Laser consolidation process for the manufacturing of structural components for advanced robotic mechatronic system—a state of art review. In: Proceedings of 6th international symposium on artificial intelligence and robotics and automation in space (i-SAIRAS 2001), Canadian Space Agency, St-Hubert, Quebec, Canada, June 18–22
39. Paul CP, Khajepour A (2008) Automated laser fabrication of cemented carbide components. Opt Laser Technol 40:735–741
40. Davis SJ, Watkins KG, Dearden G, Fearon E, Zeng J (2006) Optimum deposition parameters for the direct laser fabrication (DLF) of quasi-hollow structures. In: Proceedings of photon conference Manchester, Institute of Physics
41. He X, Yu G, Mazumder J (2010) Temperature and composition profile during double-track laser cladding of H13 tool steel. J Phys D Appl Phys 43:015502
42. Moat RJ, Pinkerton A, Li L, Withers PJ, Preuss M (2009) Crystallographic texture and microstructure of pulsed diode laser-deposited Waspaloy. Acta Mater 5:1220–1229
43. Zhong M, Liu W (2010) Laser surface cladding: the state of the art and challenges. Proc Inst Mech Eng Part C: J Mech Eng Sci 224:1041–1060
44. Valsecchi B, Previtali B, Vedani M, Vimercati G (2010) Fiber laser cladding with high content of Wc-Co based powder. Int J Mater Form 3(Suppl):11127–11130
45. Kreutz E, Backes G, Gasser A, Wissenbach K (1995) Rapid prototyping with CO2 laser radiation. Appl Surf Sci 86:310–316
46. Sun S, Durandet Y, Brandt M (2005) Parametric investigation of pulsed Nd:YAG laser cladding of Stellite 6 on stainless steel. Surf Coat Technol 194:225–231

47. Gedda H, Powell J, Wahistrom G, Li WB, Engstrom H, Magnusson C (2002) Energy redistribution during CO2 laser cladding. J Laser Appl 14:78–82
48. Draugelates U, et al (1994) Corrosion and wear protection by CO2 laser beam cladding combined with the hot wire technology. In: Proceedings of ECLAT '94, pp 344–354
49. Hensel F, Binroth C, Sepold GA (1992) Comparison of powder and wire-fed laser beam cladding. In: Proceedings of ECLAT '92, pp 39–44
50. Paul CP, Jain A, Ganesh P, Negi J, Nath AK (2006) Laser rapid manufacturing of Colmonoy components. Opt Lasers Eng 44:1096–1109
51. Paul CP, Ganesh P, Mishra SK, Bhargava P, Negi J, Nath AK (2007) Investigating laser rapid manufacturing for Inconel-625 components. Opt Laser Technol 39:800–805
52. Paul CP, Alemohammad H, Toyserkani E, Khajepour A, Corbin S (2007) Cladding of WC-12Co on low carbon steel using a pulsed Nd:YAG laser. Mater Sci Eng, A 464:170–176
53. Paul CP, Mishra SK, Premsingh CH, Bhargava P, Tiwari P, Kukreja LM (2012) Studies on laser rapid manufacturing of cross-thin-walled porous structures of Inconel 625. Int J Adv Manuf Technol 61:757–770
54. Ganesh P, Moitra A, Tiwari P, Sathyanarayanan S, Kumar H, Rai SK, Kaul R, Paul CP, Prasad RC, Kukreja LM (2010) Fracture behavior of laser-clad joint of Stellite 21 on AISI 316L stainless steel. Mater Sci Eng, A 527:3748–3756
55. http://www.utexas.edu/safety/ehs/lasers/Laser%20Safety%20Handbook-tnt.pdf. Accessed on 9 Dec 2012
56. Barat K (2009) Laser safety: tools and training. CRC Press, Boca Raton

Chapter 2
Laser Beam Machining

Shoujin Sun and Milan Brandt

Abstract The cost of cutting hard-to-machine materials by conventional mechanical machining processes is high due to the low material removal rate and short tool life, and some materials are not possible to be cut by the conventional machining process at all. Laser beam machining is the machining processes involving a laser beam as a heat source. It is a thermal process used to remove materials without mechanical engagement with workpiece material where the workpiece is heated to melting or boiling point and removed by melt ejection, vaporization, or ablation mechanisms. In contrast with a conventional machine tool, the laser radiation does not experience wear, and material removal is not dependent on its hardness but on the optical properties of the laser and the optical and thermophysical properties of the material. This chapter summarizes the up-to-date progress of laser beam machining. It presents the basics and characteristics of industrial lasers and the state-of-the-art developments in laser beam machining.

Symbols

$A_{r,\text{Fe}}$	The relative atomic mass of iron (55.8 g/mole)
B	Constant
b	Depth of focus
C_p	Heat capacity of solid metal
c_v	Volumetric specific heat of the melt
c_w	Specific heat of workpiece
D_{eff}	Diffusion coefficient of oxygen in liquid iron
d	Thickness of workpiece
E_{crit}^f	Critical energy density for the cutting fiber

S. Sun (✉) · M. Brandt
School of Aerospace, Mechanical and Manufacturing Engineering,
RMIT University, Bundoora, VIC 3083, Australia
e-mail: shoujin.sun@rmit.edu.au

M. Brandt
e-mail: milan.brandt@rmit.edu.au

J. P. Davim (ed.), *Nontraditional Machining Processes*,
DOI: 10.1007/978-1-4471-5179-1_2, © Springer-Verlag London 2013

E_{crit}^m	Critical energy density for the cutting matrix
F	Constant ($0 < F \le 1$)
f_L	Focal length of focus optics
g	Gravitational acceleration
H	Height of nozzle from the workpiece,
ΔH	Reaction heat
I	Laser beam intensity
I_0	The maximum beam intensity at r = 0 (W/cm^2)
K_m	Thermal conductivity of the melt
K_0	Bassel function of the second kind and zero order
k	Thermal diffusivity of workpiece material
L_M	Latent heat of fusion
l	Distance from the stagnation point
m_{melted}	Rate of mass gain in melt layer
$m_{ejected}$	Rate of mass loss in melt layer
P	Incident laser power
P_a	Absorbed laser energy
P_r	Power provided by the exothermal reaction in reactive laser cutting
P_{loss}	Energy loss including these to heat the solid controlled volume to the melting temperature, latent heat of melting, and heat the liquid controlled volume to cut temperature
P_v	Energy spent for vaporization, which is negligible for melting and blow process
P_0	Stagnation pressure
P_g	Average cutting gas pressure in the cut kerf
$\partial P/\partial z$	Pressure gradient through the workpiece thickness
p_a	Ambient pressure acting on the bottom of the melt film
r	Radius of the beam (cm)
r_L	Radius of laser spot on workpiece
s	Melt layer thickness
s_{ACC}	The minimum thickness of melt if melt at the bottom of cut front by acceleration of the molten material
s_{HC}	The maximum thickness of melt film without evaporation
T_0	Room temperature
T_M	Melting point of workpiece
T_E	Boiling point of workpiece
t_{on}	Pulse duration
t_{off}	Duration of laser power off in a pulse
V	Cutting speed
V_M	Velocity of cut front
v_g	Velocity of the assist gas jet
v_m	Velocity of melt flow
$\bar{v}_m$	Average velocity of melt flow
We	Weber number

w	The beam radius at which $I = I_0 e^{-2}$ (86 % of the total energy is within the beam radius w)
w_0	The collimated beam radius
w_f	Radius of focused laser beam
w_k	Kerf width
z_R	Rayleigh length distance
α	Angle of inclination of the cut front
α_w	Thermal expansion coefficient workpiece
$\hat{\alpha}$	Absorptivity of laser light
β	Cut front angles at the vertical planes
β_B	Cut front angle at the bottom
η	Percentage of melt film that is oxidized
λ	Wavelength of the laser beam
μ_g	Dynamic viscosity of the assist gas
μ_m	Viscosity of melt
θ	Angle of incidence
ρ_g	Density of the assist gas
ρ_m	Density of melt
ρ_w	Density of workpiece
σ	Surface tension of melt

2.1 Brief Introduction of Laser Beam

2.1.1 Features of Laser Beam

Laser, **L**ight **A**mplification by **S**timulated **E**mission of **R**adiation, is a high-energy beam of electromagnetic radiation. The light, photon, travels as a wave through space, but behaves as a particle of energy when it encounters matter [1]. The laser differs from other incoherent light because it is:

- monochromatic
- coherent
- directional or collimated
- bright

1. Monochromaticity

Although it is never composed of only a single wavelength in laser cavity, laser beam can be considered to be monochromatic because the oscillating laser consists of very closely spaced, discrete, and narrow spectral lines (laser modes or cavity

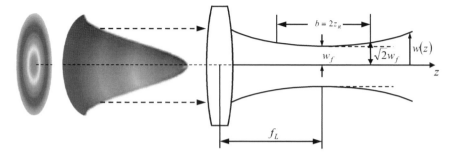

Fig. 2.1 Power distribution and focusing of TEM$_{00}$ Gaussian-mode laser beam

modes) compared with conventional light source whose emission covers frequency bandwidth in the order of gigahertz.

For better monochromaticity, single mode can be achieved by forcing laser to oscillate on a single transverse (usually, the fundamental TEM$_{00}$ Gaussian mode as shown in Fig. 2.1) and longitudinal mode.

The intensity distribution in the Gaussian beam is expressed as

$$I(r) = I_0 \exp\left(-\frac{2r^2}{w^2}\right) \tag{2.1}$$

2. Coherence

The laser beam is coherent because of the fixed-phase relationship between two waves at the wave front over time (spatial coherence) or between two points of the same wave (temporal coherence).

3. Directionality or collimation

Laser beam is directional or collimated due to its small divergence angle (Θ ranging from 0.2 to 10 milliradians except for semiconductor laser). The directionality of laser beam enables it to be focused to a very small spot over a long distance. The divergence angle (Θ) for TEM$_{00}$ beam is given as

$$\Theta \approx \frac{\lambda}{\pi \cdot w_0} \tag{2.2}$$

The beam quality is characterized by M^2 or the beam parameter product BPP (mm × mrad), given by

$$M^2 = \frac{\pi}{\lambda} w_0 \cdot \Theta \tag{2.3}$$

$$\text{BPP} = w_0 \cdot \Theta \tag{2.4}$$

Table 2.1 Characteristics of common industrial lasers for material processing [6, 153, 154]

	CO_2	Excimer	Nd:YAG	High-power diode	Fiber laser
Wavelength (μm)	10.6	0.125–0.351	1.06	0.65–0.94	1.07
BPP (mm × mrad)	12		12^a, 25–45^b	100–1,000	0.3–4
Overall efficiency (%)	5–10	1–4	1–3	30–50	10–30
Output power in CW mode	Up to 20 kW	300 W	Up to 16 kW	Up to 4 kW	Up to 10 kW
Focused power density (W/cm^2)	10^{6-8}		10^{5-7a}, 10^{6-9b}	10^{3-5}	
Pulse duration (sec)	10^{-4}	10^{-9}	10^{-8}–10^{-3}	10^{-12}	10^{-13}
Fiber coupling	No	Yes	Yes	Yes	Yes

^{a} Pumped by flash lamp
^{b} Pumped by diode

4. Brightness

Brightness measures the capability of a laser oscillator to emit a high optical power per unit area per unit solid angle, which is related to the directionality of the laser beam. Smaller divergence angle results in higher brightness. Laser beam machining process requires a laser beam with high brightness (BPP is generally smaller than 20 at power level of kW).

Lasers are normally classified based on the physical states of the laser-active (pumping) medium as gas lasers (CO_2 and excimer), solid state lasers (Nd:YAG), semiconductor lasers (diode), liquid lasers, and fiber lasers (Yb:YAG). Characteristics of the common industrial lasers are listed in Table 2.1.

The excimer laser may be characterized by its intense UV-wavelength pulses which remove material by unique photochemical ablation process [2]. The excimer laser is normally used to process organic materials.

Despite their high efficiency, diode lasers have low beam quality with high beam parameter product, which limits these lasers being applied in cutting, drilling, marking on metallic materials, and micromachining [3]. However, high-power diode lasers are well suitable for preheating workpiece for laser-assisted machining for example.

The most important characteristics of high-power laser for cutting are [4] as follows:

1. The beam must as close as possible to the fundamental TEM_{00} radial mode;
2. Both output power and its radial distribution must be temporally constant in the continuous wave mode.

2.1.2 Manipulation of Laser Beam

Laser beam can be manipulated through optics such as follows:

Focusing: Laser beam can be focused by curved mirrors or lenses to reduce its spot size. The radius of focused Gaussian-mode laser beam (w_f) is the smallest laser beam radius at the plane $z = 0$ as shown in Fig. 2.1:

$$w_f = \frac{f_L \cdot \lambda}{\pi \cdot w_0} M^2 = f_L \cdot \Theta \cdot$$ (2.5)

The depth of focus (b) is twice the Rayleigh length distance z_R, which is defined as the distance from the focus at which the cross-sectional area of the laser beam has doubled (or the beam radius increases by a factor of $\sqrt{2}$).

$$b = 2z_R = \frac{2\pi \cdot w_f^2}{\lambda \cdot M^2}$$ (2.6)

It indicates in Eqs. (2.5) and (2.6) that both the focused radius and depth of focus of a laser beam with Gaussian mode increase with the focal length of focusing optics.

Shaping (spatial shaping): The commonly used laser beam for cutting purpose is the Gaussian beam (TEM$_{00}$), but it can be shaped into other profiles such as top-hat and rectangular beams, which have a complex power distributions rather than a Gaussian distribution as shown in Fig. 2.1.

Pulsing (temporal shaping): The laser beam can be emitted temporally in either the continuous wave (CW) or pulse modes. The power of laser beam is constant in the continuous wave mode, but intermittent in the pulse mode. The pulse laser beam is characterized with peak power, pulse duration (length or width), and pulse frequency (repetition frequency). The advantage of the pulsed laser is its increasing peak power with shorter pulse duration without an increase in the average output (except in the case of gated pulsing). From the processing point of view, the depth of heat conduction (heat-affected zone) into workpiece is lower compared with that of CW laser [5]. Laser pulse shape can be manipulated to generate a rectangular, smooth, and triangular pulses.

Pulsed laser beam can be classified into three categories [6]:

1. Gated pulsing: peak power is about the same with nominal output power in CW mode
2. Superpulsing: The peak power is higher than the average output power depending on the pulse frequency and duty cycle.
3. Hyperpulsing: High peak power pulses are superimposed on the CW output power.

The pulsing is achieved by normal pulsing, Q switching (generation of ns pulses), and mode locking (for the generation of pulses with pulse widths below 100 picoseconds down to a few femtoseconds).

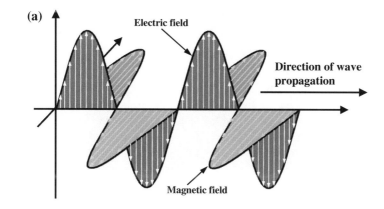

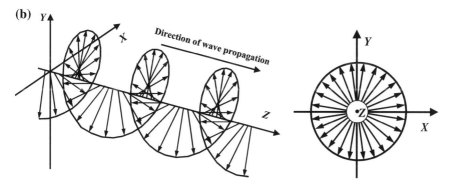

Fig. 2.2 Electric field vector in (**a**) linearly and (**b**) circularly polarized beams

Polarization: laser beam emitted from the laser cavity can be linearly polar-ized, that is, the vectors of the oscillating electric and magnetic field of the traveling photons are perpendicular to each other and to the propagating direction; the electric field vector oscillates only in one plane (as shown in Fig. 2.2a).

In order to eliminate the effect of orientation of electric field vector, the laser beam can be depolarized through optical device (depolarizer) as a circularly polarized beam in which the direction of the electric field vector rotates about its propagation direction while it retains constant magnitude. The tip of the electric field vector, at a given point in space, describes a circle as time progresses (as shown in Fig. 2.2b).

Splitting: A laser beam can be split into multiple beams with various power ratios between beams by using an optical device (beam splitter) through the methods of reflection, interference, or polarization.

Spinning: A focused beam is shifted laterally by rotating mirror or window to cover wider area.

2.2 Characteristics of Laser: Material Interaction

2.2.1 Photochemical (Photolytic) Process

When a material (normally an organic material) is irradiated by a short-wavelength (in UV range, such as excimer laser) and short-pulse-length (shorter than the thermal relaxation time in microseconds) laser beam, the molecular bonds in a very thin layer of the material surface can be broken by the higher energy of shorter-wavelength photon with minimum thermal effect, and this is also called "cold-cutting." The photon energies between 3.60 and 4.29 eV within the wavelength between 344 and 288 nm are high enough to break C–C and C–H covalent bonds (with average binding energies of 347 kJ/mol and 414 kJ/mol, respectively) [7].

2.2.2 Photothermal (Pyrolytic) Process

When a laser beam irradiates the workpiece surface, part of the laser beam energy is absorbed by the workpiece surface due to the interaction between the electromagnetic radiation and electrons of the workpiece materials. The remaining is lost due to the reflection by the workpiece surface.

The absorption rate, or absorptivity, characterizes the percentage of incident laser power that is absorbed by a solid workpiece material. It is affected by the wavelength of laser beam (as shown in Fig. 2.3), polarization of beam, characteristics of workpiece material (chemical composition, surface condition, and temperature), and beam incident angle.

The laser beam absorption increases with surface roughness only when it exceeds a roughness threshold, which is dependent on the type of workpiece

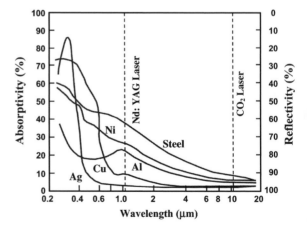

Fig. 2.3 Absorption and reflection rates as a function of beam wavelength for various metals (drawn in reference to [146])

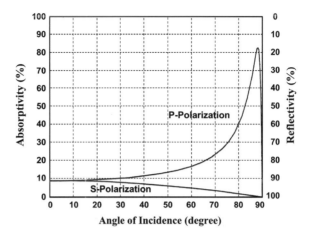

Fig. 2.4 Variation of reflectivity and absorptivity of iron to the linearly polarized CO_2 laser beam with angel of incidence (drawn in reference to [147])

materials, and the relationship between absorption and surface roughness is affected by the beam incident angle [8].

The beam incident angle has a strong effect on the absorption of linearly polarized beam, which is determined by Fresnel absorption [6, 9]. Generally, the refection of s-polarized beam (the electric field vector is normal to the plane of incidence) is higher than that of the p-polarized beam (the electric field vector is in the plane of incidence) as shown in Fig. 2.4. Brewster's angle at which the absorptivity of the incident beam reaches maximum is very high and near 90°.

The dependence of absorptivity on angle of incidence for unpolarized laser beam is strongly affected by the beam wavelength as shown in Fig. 2.5. The Brewster's angle is 79.6° and 87.3° for fiber laser with wavelength of 1.07 μm and CO_2 laser with wavelength of 10.6 μm, respectively. The absorptivity of molten

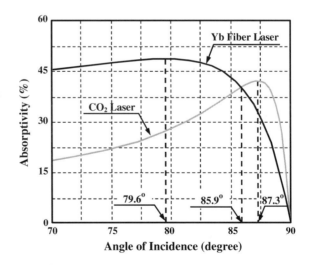

Fig. 2.5 Effect of angle of incidence on absorptivity of molten iron for unpolarized laser beams with different wavelengths (reprinted from Mahrle and Beyer [31] with permission. Copyright © IOP Publishing Ltd)

iron for fiber laser radiation is better than that for CO_2 laser radiation only when the angle of incidence is smaller than 85.9°.

Surface directionality due to the mechanical surface finish is also found to affect the absorption of polarized light [10].

The absorbed laser energy is converted to the thermal energy to cause a significant temperature rise on the surface of the workpiece, which can

1. induce thermal stress in the heat-affected zone if the surface temperature is lower than the melting point;
2. melt the surface of material when the surface temperature is higher than the melting point, but lower than the boiling point;
3. vaporize the surface of material when the surface temperature is higher than the boiling point.

Vaporization occurs on the material surface when the surface temperature reaches its boiling point with high laser intensity ($>10^8$ W/cm^2), which is usually achieved with pulsed laser (superpulsing or hyperpulsing). Recoil pressure is generated on the melt surface due to the vapor evolution, which can push the melt layer from the molten zone when the tangential recoil pressure exceeds the surface tension [11]. Because the material surface is partially melted and vaporized, the material removal process without assist gas comprises both melt expulsion due to the recoil pressure and vaporization [12].

A plasma plume may be formed at or near to the surface of the materials as the result of interaction between laser beam and vapor. This plasma plume would absorb most of the incident beam energy and lead to decoupling of the incident beam from the workpiece surface [13], thereby lowering the process efficiency [14]. The proportion of incident laser power absorbed by the plasma is lower with a shorter wavelength of laser. Hence, the material removal rate is higher for shorter-wavelength laser radiation during ablation [2].

2.3 Laser Beam Machining

2.3.1 Integration of Laser System for Cutting Operation

In order to process workpiece, the laser beam has to be delivered and integrated with a motion control system. This system comprises the following:

1. Beam delivery system

Due to its long wavelength, CO_2 laser cannot be transmitted through optical fibers. Therefore, it can only be delivered to the workpiece by a series of metallic mirrors, whereas the other industrial lasers due to their shorter wavelength can be delivered by optical fiber, which offers higher degree of freedom and flexibility in a manufacturing environment.

2. Cutting head

It consists of beam delivery optics, focusing lens and gas delivery nozzle, which is usually assembled coaxially with laser beam. Two jets of gas are possibly delivered by the nozzle: one to the laser spot to remove the melted metal and the other to the surrounding of laser beam to prevent oxidation as shielding gas. The head is positioned to ensure that the laser beam is incident perpendicular to the surface of the workpiece. The gap between the nozzle and surface of workpiece must be maintained constant and at a specified stand-off distance.

3. Workpiece positioning

The relative motion between laser beam and workpiece can be achieved by either moving the workpiece only, or moving cutting head only or moving both cutting head and workpiece (hybrid system).

Flying optics: the relative movement of the laser spot can be achieved by moving only mirrors and the focusing lens while the workpiece remains stationery for remote cutting without gas jet, which allows high positioning and processing speeds with high precision and change in spot size.

2.3.2 Laser Beam Machining Processes

Some laser beam machining processes are summarized in Table 2.2. In photo-thermal processes, material can be machined by the following:

(1) Controlled fracture process: brittle materials (ceramics) are separated by thermally induced stress cracking, which requires very low energy. Cutting speed or thickness of workpiece can be increased by employing dual-beam CO_2 laser [15] or two lasers [16]. It can be done in air or with water jet quenching after laser radiation [17]. The cutting speed with water assistance is reported to be 500 % faster than that in dry machining [18]. However, crack initiation and propagation at corners and curves during profile cutting is difficult to control and leads to cut deviation [19].

(2) Melting and blow process: The surface of workpiece is heated and melted by laser beam (and exothermic heat if reactive gas is used) and forms an erosion/cutting front (a layer of molten material). The erosion/cut front that extends through the thickness of the workpiece is subsequently blown away by a pressurized gas jet through a coaxial nozzle (positioned coaxially with the laser beam) or off-axis nozzle as shown in Fig. 2.6. It is classified as laser fusion cutting in the case of inert gas and reactive laser cutting when a chemically reactive gas is used.

The CO_2 and Nd:YAG lasers are the two industrial lasers used for decades for removing material by melting and evaporation. Despite its lower average power

Table 2.2 Summary of laser beam machining processes [155]

	Controlled fracture	Photo-thermal processes		Photo-chemical process
		Melting and blow	Vaporization	
		Fusion	Reactive		
Mechanism	Material is broken as the result of thermally induced stress	Material is melted as a result of laser radiation, and the melt is blown by pressurized gas jet	Material is melted as a result of laser radiation and exothermic energy, the melt is blown by pressurized gas jet	Material is vaporized as a result of high intensity of laser beam radiation	Material is separated as the result of broken covalent bonds by photon energy of UV light
Relative laser energy required	1	20	10	40	100
Gas type	Inert gas	Inert gas	Chemically reactive gas	Inert gas or no gas	Inert gas
Workpiece materials	Ceramics	Metals, polymers, ceramics, glasses, composites	Metals, polymers	Metals, ceramics, polymers	Organic materials

Fig. 2.6 Illustration of laser beam cutting process (drawn in reference to [148])

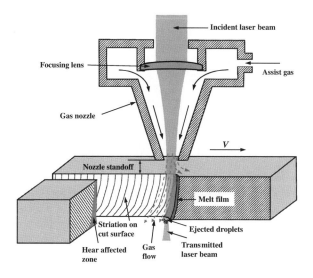

compared with CO_2 laser (for cutting thick material >10 mm), the Nd:YAG laser offers relatively high peak power in the pulse mode due to its shorter pulse duration (as shown in Table 2.1) and better focusing properties and therefore the better cut quality in terms of kerf width, heat-affected zone (HAZ), and cut edge quality. Due to its smaller thermal load, Nd:YAG laser can machine brittle ceramic materials which CO_2 laser cannot machine without cracks [20]. Due to its shorter wavelength, pulsed Nd:YAG laser can be absorbed more effective by the highly reflective metallic workpiece as shown in Fig. 2.3. Another advantage of Nd:YAG lasers is their transportability by optical fiber, which makes the cutting system more flexible and output beam non-polarized.

(3) Evaporation (sublimation cutting) process:

The material is removed by vaporization and the recoil pressure generated as vapor evolution as the surface temperature reaches the boiling point. The energy required for evaporation is estimated to be 6 times the energy required for fusion cutting [21].

This chapter is concentrated on the laser beam machining with material removed by melting and ejection because of its wide applicability to cut a large variety of materials. Some examples involving partial vaporization are also briefly discussed.

2.3.2.1 Process Analysis

The top and side views of a laser cutting process are shown in Fig. 2.7. A layer of melt at an angle of inclination (α) is produced through the thickness of workpiece as a result of heat input from laser beam in laser fusion cutting or laser power plus

Fig. 2.7 a Top view and
b side view of cut front
(drawn in reference to [29,
48, 61, 149]). Note the
different scales in (a) and (b)

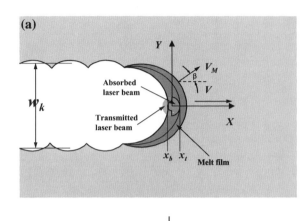

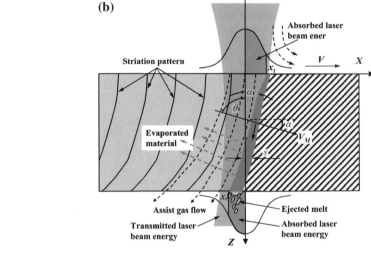

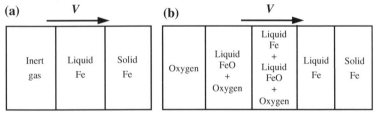

Fig. 2.8 Boundaries in the cut front for (**a**) laser fusion cutting and (**b**) oxygen-assisted laser cutting of mild steel (drawn in reference to [39])

exothermic heat in reactive laser cutting. A pressurized gas jet is used to eject the melt out of the kerf in the direction of cut front plane in order to keep the process ongoing when the laser beam is moving relative to the workpiece. Vaporization and melt expulsion by recoil pressure can contribute to the melt ejection at cutting temperature above the boiling point as in laser fusion cutting and remote cutting process, the material lost by evaporation due to recoil pressure is predominantly in the direction perpendicular to the cut front [22].

The boundaries in the cut front are schematically shown in Fig. 2.8 for both laser fusion cutting and oxygen-assisted laser cutting of mild steel.

1. The melt front: the boundary between solid and liquid moves depending on the heat balance at the boundary [23].
2. The melt surface: the boundary between liquid and gas moves depending on the heat balance affected by laser beam absorption, melt flow, and heat conduction [23].
3. The oxidation front in reactive laser cutting: the boundary between liquid Fe and oxide layer moves dynamically at the velocity determined by diffusion of oxygen through oxide layer [24–27] or by the diffusion of iron through oxide layer [28].

The laser cutting process involves the following balances as listed in Table 2.3 for both the steady state and dynamic state. The temperature is denoted as T and thickness s of the melt fluctuate in the dynamic state.

The energy input includes the following:

1. the absorbed laser power at the cut front

Only partial laser beam is incident on the cut front (from x_b at the bottom surface to x_t at the top surface as shown in Fig. 2.7). Therefore, part of laser power is lost due to transmission through the kerf when the beam is outside the x_b. Hence, increase in the horizontal lag between the top and bottom of cut front $(x_t - x_b)$ enables more laser energy to be absorbed at the cut front. The absorbed laser energy is written as [29]

$$P_a = \hat{\alpha} \int_0^{xt} I(r, z = 0) r\pi \, \mathrm{d}r + \hat{\alpha} \int_{xb}^0 I(r, z = \mathrm{d}) r\pi \, \mathrm{d}r \qquad (2.7)$$

Table 2.3 List of balances at the cut front

	Steady state [29]	Dynamic state [49]
Mass balance	$m_{\text{melted}} = m_{\text{ejected}}$	$m_{\text{melted}} - m_{\text{ejected}} = \rho_m \cdot d \cdot w_k \cdot \frac{ds}{dt}$
Energy balance	$P_a + P_r = P_{\text{loss}} + P_v$	$P_a + P_r - P_{\text{loss}} + P_v$
		$= c_v \cdot d \cdot w_k \cdot s \cdot \dfrac{dT}{dt} + c_v \cdot d \cdot w_k \cdot T \cdot \dfrac{ds}{dt}$
Momentum balance	$F_0 + F_n + F_t = F_a + F_{\text{st}} + F_d + F_m$	

P_{loss}, the energy losses including those to heat the solid controlled volume to the melting temperature, latent heat of fusion, and heat the liquid controlled volume to cut temperature, and power loss by heat conduction to the surrounding solid workpiece. The contributions of heat loss by convection and radiation are negligible compared with conduction loss [30, 32, 39, 50], P_v, the energy spent for vaporization, which is negligible for the melting and ejection process

The absorptivity $\hat{\alpha}$ increases with workpiece thickness [21, 30]. This is the result of change in the angle of incidence due to change in inclination angle of the cutting front [31, 32]. The maximum absorptivity of iron ($\hat{\alpha} = 0.41$) is achieved at the angle of incidence of $\theta = 87.3°$ (the corresponding angle of inclination of cut front is $\alpha = 2.7°$) for the unpolarized CO_2 laser beam as shown in Fig. 2.5. The increasing effect of multiple reflections with an increase in workpiece thickness also enhances the beam absorption in the kerf [33].

2. the exothermic heat in reactive laser cutting

When oxygen is used as assist gas for cutting mild steel, significant heat is generated during the oxidation of iron as

$$Fe + \frac{1}{2}O_2 = FeO, \Delta H = -257.58 \text{kJ/mol (at 2,000K)} \tag{2.8}$$

The energy contributed by exothermic reaction becomes saturated with increase in pressure of gas flow (the number of pure metal and oxygen particles in the molten region) [34]. It is calculated as [29, 30]

$$P_r = \frac{w_k \cdot \rho_m \sqrt{2D_{\text{eff}} \cdot d \cdot v_m}}{A_{r,\text{Fe}}} \Delta H \tag{2.9}$$

$$P_r = \frac{\eta \cdot d \cdot w_k \cdot V \cdot \rho_m}{A_{r,\text{Fe}}} \Delta H \tag{2.10}$$

The contribution of exothermic heat to the total cutting energy is about 40 % [35–37], 55–70 % [38], and 50–56 % and is independent of the workpiece thickness calculated from [30] for cutting mild steels and 60 % for cutting stainless steel [37]. Both the exothermic heat and absorbed laser energy increase with cutting speed, but the percentage of exothermic heat to the total cutting energy decreases with increase in cutting speed [26]. Therefore, the cutting process is significantly enhanced in terms of lower laser power, low gas pressure, and higher cutting speed with oxygen as assist gas compared with nitrogen as assist gas [39]. Application of oxygen as assist gas allows cutting speeds that are 3–6 times higher than those in laser fusion cutting with inert gas [40, 41].

Not all the molten iron in the cutting front is oxidized [39], the percentage of melt film that is oxidized increases with decrease in cutting speed due to the thinner melt film and longer reaction time at low cutting speed. Only 30–35 % of iron that undergoes oxidation contributes to the total energy balance of laser cutting. The degree of oxidation is independent of the workpiece thickness [30]. But Ivarson et al. [37] believed 50 % of molten iron when cutting mild steel, 30 % of iron and 30 % chromium when cutting stainless steel are oxidized, and the majority of products of the exothermic reaction are FeO during cutting of mild steel, Fe_2O_3, and Cr_2O_3 when cutting stainless steel. The oxidation levels in the particles ejected from different parts of cut front are different because of the difference in their exposure times to oxygen (stages of oxidation) [42]. The oxide melt, FeO, when cutting mild steel, is of low viscosity and low surface tension at temperature above 2,000 K and therefore is easily removed from the cutting zone by the mechanical action of the oxygen jet [24, 39], but the Cr_2O_3 formed when cutting stainless steel is of high melting point (>2,700 K), which increases the viscosity of melt and hinders the diffusion of oxygen into melt [36].

In the momentum balance,

F_o, the gas static pressure [29],

$$F_o = \frac{\pi}{2} d \cdot w_k \cdot p_g \tag{2.11}$$

F_n, the normal component of dynamic gas force [29],

$$F_n = \frac{\pi}{2} d \cdot w_k \cdot p_g \cdot v_g \tan \alpha \tag{2.12}$$

F_t, the tangential component of dynamic gas force [29],

$$F_o = 2\pi w_k \sqrt{d \cdot \rho_g \cdot \mu_g \cdot v_g^3} \tag{2.13}$$

F_a, the force acting on melt due to ambient pressure [29],

$$F_a = s \cdot w_k \cdot p_a \tag{2.14}$$

F_{st}, the surface tension force of melt [29],

$$F_{st} = \pi d \cdot \sigma \tag{2.15}$$

F_d, the dynamic force [29],

$$F_d = \frac{\pi}{6} w_k \cdot \rho_m \cdot v_m^2 \cdot s \tag{2.16}$$

F_m, the force due to the friction loss in the melt [29],

$$F_m = \frac{\pi}{2s} w_k \cdot d \cdot \mu_m \cdot v_m \tag{2.17}$$

2.3.2.2 Characteristics of Cut Front

Laser cutting is the process of formation and removal of the cut front. Therefore, its characteristics, such as temperature, angle of inclination, and flow rate, determine the cut quality. Hence, the effect of processing parameters on these characteristic must be fully understood.

Considerable efforts have been made to investigate the characteristics of cut front by in-process visualization and monitoring [25, 43–47], and theoretical and modeling analysis [29, 38, 48–54].

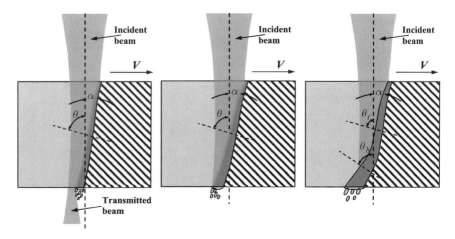

Fig. 2.9 Change in cut front with increase in cutting speed from (**a**) to (**c**)

Angle of Inclination of the Cut Front

It is found that the angle of inclination of the cut front increases with increase in cutting speed for both laser fusion cutting and reactive laser cutting [23, 29, 33, 45, 55], and the cut front becomes curved or kinked at high speeds [56] as illustrated in Fig. 2.9. The curved cut front at the lower part (Fig. 2.9c) with larger angle of inclination at high cutting speed is due to the lower laser irradiance in the lower part of the cut front than that in the upper part when the laser beam is focused close to the top surface of the workpiece [46] and will result in a change in melt flow direction [56].

The fraction of the laser beam incident on the cut front increases as the result of increase in angle of inclination of cut front at higher cutting speed.

In laser fusion cutting, the inclination angle increases with decrease in the linear energy density (either increase in cutting speed or decrease in laser power). An increase in the assist gas pressure increases the inclination angle probably due to the consequence of energy losses of higher removal of molten material [46].

Due to the increase in the Rayleigh length with shorter wavelengths for a Gaussian beam with $M^2 = 1$ as shown in Eq. (2.6), the angle of inclination of cut front in the case of fiber laser ($\lambda = 1.07\,\mu m$) fusion cutting is larger than that of CO_2 laser beam fusion cutting, and the difference increases with increase in workpiece thickness [31].

The angle of inclination of the cut front reduces locally due to the formation of "shelves" on which melt accumulates and forms humps (as shown in Fig. 2.10). The accumulations of melt are produced periodically leading to melt flow instability [45].

In the remote fusion cutting without assist gas, the cutting front angle (complementary angle to angle of inclination α) decreases with increase in cutting speed and decrease in laser power, but is independent of the workpiece thickness.

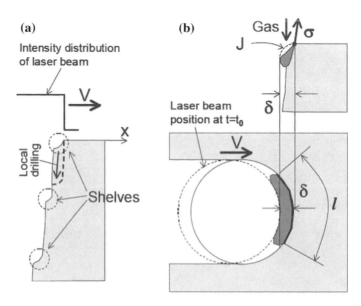

(a)

Intensity distribution
of laser beam

V →

X →

Local drilling

Shelves

(b)

Gas ↓↑ σ

J

Laser beam
position at t=t₀

δ

V →

δ

l

Fig. 2.10 **a** Side view and **b** top view of formation shelves and accumulation of melt (humps) (reprinted from Hirano and Fabbro [55] with permission. Copyright © Elsevier)

The maximum and the minimum cutting speeds are achieved at the minimum and the maximum cutting front angle, respectively, which are dependent on workpiece thickness [57].

The ratio of the area of cut front that is directly irradiated by the laser beam to the total area of cut front decreases with increase in the angle of inclination of the cut front after the full laser beam is incident on the cut front. The stable cut is still possible with the cut front not completely covered by the laser beam [56, 57].

If the angle of incidence is far from the Brewster's angle due to the change in angle of inclination of the cut front with increase in workpiece thickness and cutting speed, multiple reflections become significant and affect the cut process [33].

Temperature at Cut Front

In order to keep the cutting processing continuous, the temperature at the cut front has to be maintained above the material's melting point. The temperature at cut front is measured at approximately 2,000 K and maintained below boiling point of iron for reactive laser cutting of mild steel. FeO will not form (or, having formed, it will dissociate) if temperature in the melt exceeds the dissociation temperature (3,600 K), which makes the exothermic energy from the oxidation reaction not available as an energy input to the cut zone [39]. However, some modeling analyses show the melt temperature well above the boiling temperature [28, 38, 58–60].

Cut front temperature increases with laser power [48, 61] and cutting speed for both laser fusion cutting [29, 45] and reactive laser cutting [25, 39, 58, 62, 63]. This could be due to the following:

1. the greater beam coupling to the work material as the fraction of laser beam transmitted straight through the kerf with no heating effect decreases because of the increase in angle of inclination of the cut front [58];
2. increase in beam absorptivity with decrease in angle of incidence (increase in angle of inclination of the cut front) before reaching the Brewster's angle at higher cutting speed;
3. increase in exothermic heat in reactive laser cutting [26];
4. proportion of energy loss from cutting zone decreases with increase in cutting speed [26, 64].

The temperature of the melt surface is not uniform through the thickness of workpiece. Higher temperature is achieved at the region closer to the bottom surface [62]. The melt surface temperature reaches the maximum temperature at the maximum cutting speed. The process breaks down at cutting speed higher than the maximum cutting speed because of abrupt drop in temperature of cut front at the region near the bottom surface [62].

It is reported that cut front temperature decreases with increase in workpiece thickness [48, 61]. The maximum thickness of a workpiece that can be cut at a given laser power is when the cut front temperature reaches the melting point as shown in Fig. 2.11 [48].

Cut front temperature is also found to be affected by beam polarization. It is higher with circularly polarized beam than with the radially polarized beam regardless of cutting speed [65].

It should be noted that the temperature of the melt surface fluctuates in reactive laser cutting because of oxidation dynamics [26].

Fig. 2.11 Effect of workpiece thickness on the cut front temperature (reprinted from Schuöcker and Abel [48] with permission. Copyright © SPIE)

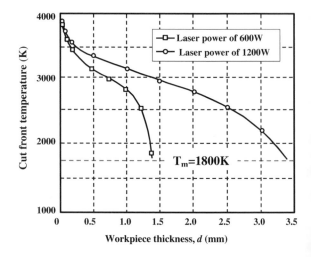

The Melt Film Thickness

The melt film is very thin (of order of 10^{-5} m), and its thickness generally increased from the top to the bottom of the kerf [62, 66] as the melt accumulates at the bottom of cut front.

The melt film thickness, s, is formulated by Wandera et al. [54] for laser fusion cutting as

$$s = \frac{24V \cdot d^2 \left(\mu_g + \mu_m\right)}{\rho_g \cdot v_g^2 \cdot w_k^2} \tag{2.18}$$

and by Schuöcker et al. [34] for laser reactive cutting as

$$s = \frac{2k}{V} \left(1 - \frac{T_s}{T}\right) \frac{K_0 \left(Vw_k/4k\right)}{K_1 \left(Vw_k/4k\right)} \tag{2.19}$$

It clearly shows that the melt layer thickness increases with

- increase in thickness of workpiece as shown in Fig. 2.12 [48];
- increase in cutting speed [53, 56, 62, 67, 68] since more material is melted per unit time during cutting and higher mass flow rate at higher cutting speed, the thickness will be distorted at the region near the bottom surface when the cutting speed is close to the maximum cutting speed due to the inertial forces [62];
- decrease in gas velocity in laser fusion cutting [53].

However, Kaplan's analysis shows that the melt film thickness decreases with increase in workpiece thickness at a given cutting speed and laser power for laser fusion cutting as shown in Fig. 2.13a [29].

In reactive laser cutting of mild steel, a layer of FeO is formed on the surface of melt film, its thickness decreases with increase in cutting speed [69] and pressure

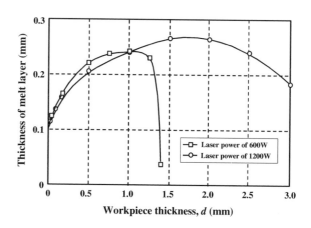

Fig. 2.12 Effect of laser power and workpiece thickness on melt layer thickness (reprinted from Schuöcker and Abel [48] with permission. Copyright © SPIE)

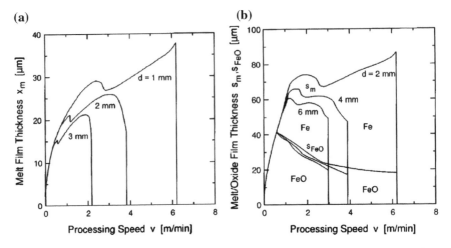

Fig. 2.13 Melt film thickness as function of cutting speed and workpiece thickness in (**a**) laser fusion cutting and (**b**) reactive laser cutting (reprinted from Kaplan [29] with permission. Copyright © American Institute of Physics)

of O_2 jet, but the fraction of oxide layer decreases with increase in cutting speed (as shown in Fig. 2.13b) and increases with increase in O_2 jet pressure [29].

The thickness of the melt film in reactive laser cutting fluctuates with the same pattern of the fluctuation of melt surface temperature [26].

The boundary layer thickness can be increased if shock structure in the assist gas flow field is formed inside the kerf because of the reduction in the flow velocity [70, 71].

In order to investigate the cutting conditions without evaporation, Olsen calculated and compared two melt film thicknesses, the maximum thickness of melt film in laminar one-dimensional melt flow (s_{HC}) and the minimum thickness of melt at the bottom of cut front by acceleration of the molten material (s_{ACC}) as [66]

$$s_{HC} = \frac{(T_E - T_M)K_m}{\left((T_M - T_0)C_p + L_M\right) \cdot \rho_m \cdot V \cdot \beta_B} \qquad (2.20)$$

$$s_{ACC} = V \cdot d \sqrt{\frac{\rho_m}{P_g}} \qquad (2.21)$$

Cutting without evaporation occurs when

$$s_{HC} > s_{ACC} \qquad (2.22)$$

Thickness s_{HC} reduces with increase in cutting speed whereas s_{ACC} increases with increase in cutting speed and workpiece thickness, but decreases with increase in gas pressure as shown in Fig. 2.14.

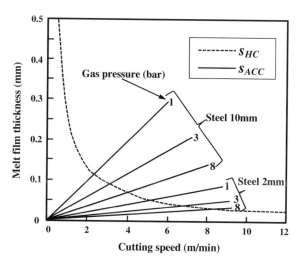

Fig. 2.14 Melt film thicknesses of s_{HC} and s_{ACC} as function of cutting speed, gas pressure, and workpiece thickness (reprinted from Olsen [66] with permission. Copyright © SPIE)

Melt Flow Velocity

When the highest melt surface temperature is lower than the boiling temperature, the melt flows downstream in one dimension due to two driving forces acting on it by the momentum transfer of the pressurized gas flow in laser fusion cutting with high-pressure gas jet [50]:

1. pressure gradient of the gas flow
2. shear stress due to viscous friction within the boundary layer

Both mechanisms tend to remove the molten material in a direction parallel to the cutting front, and both contributions are of the same order of magnitude and increase with increase in gas velocity and angle of inclination of the cutting front [50].

However, the gas pressure is much lower in reactive laser cutting, the pressure gradient of the gas flow is negligible. The melt is considered to be removed primarily by shear stress due to the viscous friction [72, 73]

The maximum melt flow velocity, v_m, at the gas/melt interface is found in laser fusion cutting [54]:

$$v_m = \left(\frac{1}{\mu_g + \mu_m}\right)\left(\frac{\rho_g \cdot v_g^2}{16d}\right) w_k^2 \qquad (2.23)$$

in laser fusion cutting [74]:

$$v_m = \frac{\rho_g \cdot l}{24\mu_m}\left(\frac{F \cdot v_g}{H}\right)^2 w_k \qquad (2.24)$$

and in reactive laser cutting [48, 61]:

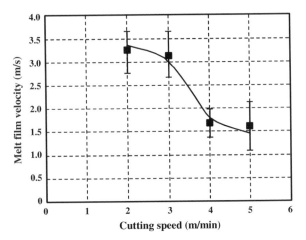

Fig. 2.15 Velocity of melt film flow as a function of cutting speed in laser fusion cutting (reprinted partially from Hirano and Fabbro [45] with permission. Copyright © IOP Publishing Ltd)

$$v_m = \sqrt{\frac{\mu_g}{\rho_m} \frac{d}{s \cdot w_k}} v_g \qquad (2.25)$$

From the above equations, it can be concluded that high velocity and density of gas is required to increase the melt flow velocity.

Increase in cutting speed results in increase in cut front temperature (reduction in viscosity of the melt), angle of inclination of cut front and melt film thickness, and slight reduction in kerf width [24, 29]. The combination of these effects leads to the reduction in melt flow velocity as show in Fig. 2.15 [45]. However, Chen et al. shows that the melt surface speed increases with cutting speed [67].

When the highest surface temperature (at the bottom part of front) exceeds the boiling temperature with increase in cutting speed, evaporation takes place, which

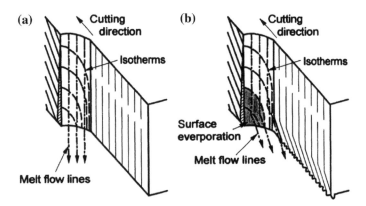

Fig. 2.16 Contours of isothermals and melt flows in (**a**) one-dimensional flow without evaporation and (**b**) two-dimensional flow with evaporation (reprinted from Olsen [66] with permission. Copyright © SPIE)

results in a local decrease in the melt thickness. If the evaporation is substantial, that is, when the evaporation pressure exceeds the cutting gas pressure locally, the melt flow will be forced not only downwards in the cut kerf (one-dimensional) but also around the laser beam (two-dimensional) as shown in Fig. 2.16 [66].

Mobility of Cut Front

Cut front does not always move at the same velocity with laser beam. It could move in front of or behind the laser spot depending on the heat balance affected by laser beam absorption, exothermic reaction, melt flow, and heat conduction. In the steady-state cutting, the melt front at a point moves at a velocity of V_M depending upon the angle between the cutting direction and the local normal to the melt front as [66]

$$V_M = V \cos \alpha \cdot \cos \beta \qquad (2.26)$$

V_M is proportional to the corresponding local temperature gradients in the melt. The higher the V_M, the steeper temperature gradient is needed in the melt [66].

Difference in velocities of cut front and laser beam is also found during acceleration and deceleration in profile cutting, which is related to the issue of non-linear beam coupling to the workpiece [58].

Separation of Boundary Layer Flow

The melt velocity profile is dependent on the external pressure gradient through the workpiece thickness $\left(\frac{\partial P}{\partial z} \right)$ [54]:

- when $\frac{\partial P}{\partial z} = 0$, the velocity gradient gradually reduces to 0 at the outer edge of the boundary layer. The boundary layer is a perfect laminar flow;

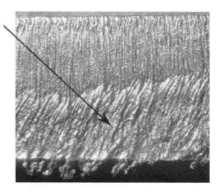

Fig. 2.17 Poor surface quality at lower region of cut surface due to flow separation (reprinted from Kovalev et al. [75] with permission. Copyright © Elsevier)

Flow separation zone

- when $\partial P/\partial z$ increases to an extreme adverse pressure gradient $\partial P/\partial z \gg 0$, the velocity profile is significantly distorted at the kerf wall. The separation of flow occurs at the point of velocity gradient of zero, from which the boundary layer flow transitions into turbulent flow. The shear stress resulting from friction between the gas stream and melt is reduced significantly at the separation point [68].

When the depth of flow separation is smaller than the thickness of workpiece, rough surface finish at the lower region and strong dross attachment to the lower cut surface are produced as shown in Fig. 2.17.

In order to maintain a high cutting quality, a laminar boundary layer melt flow must be maintained throughout the cut depth. Therefore, it is desired to have a depth of flow separation being larger or equal to the workpiece thickness to obtain clear cut. The depth of flow separation can be increased in laser fusion cutting [54] by:

1. increasing the gas pressure to reduce melt film thickness [75];
2. increasing nozzle diameter;
3. focusing the laser beam close to the bottom edge of the workpiece for thick workpiece with sufficient power intensity;
4. increasing melt surface temperature by increasing cutting speed and laser power;
5. increase in the kerf width [68].

2.3.2.3 Characteristics of Cut Surface

The characteristics of cut surface are important because the quality of the subsequent operations (such as welding) is strongly affected by the characteristics of cut surface, which are described as striation, dross as shown in Fig. 2.18, and heat-affected zone.

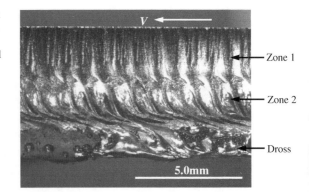

Fig. 2.18 A typical poor cut surface showing striations in zones 1 and 2 and dross produced in cutting mild steel by reactive laser cutting process

Striation

Striations are regularly periodic lines produced on cut surface (as shown in Fig. 2.18) during both laser fusion cutting and reactive laser cutting processes. The angle of the striations relative to the laser beam axis increases with increase in cutting speed.

Two different patterns of striation are observed on the cut surface when workpiece thickness is over 2 mm [24]. One is near the top surface adjacent to the cutting head with a relatively fine pattern (short wavelength) and another is at the lower part of the cut edge with a relatively coarse pattern (much longer wavelength). The two patterns are separated by a straight line extending in parallel to the surface of the workpiece [49]. The separation line is shifted toward the top surface of the sheet metal with increase in cutting speed [43].

The subsurface in the striation at zone 1 does not show resolidified microstructures [24, 43], but the striation at zone 2 shows the resolidified microstructures in its subsurface, which indicates that the striation at zone 2 is formed as a result of the turbulent melt film resolidification on the cut surface before it is ejected from the kerf.

Deep striation affects the quality of the cut surface (high surface roughness, geometry accuracy, etc.). Striation is characterized by its depth (surface roughness) and wavelength. It is found that striation depth (surface roughness) increases with increase in workpiece thickness [76] and oxygen jet pressure [26, 67, 77]

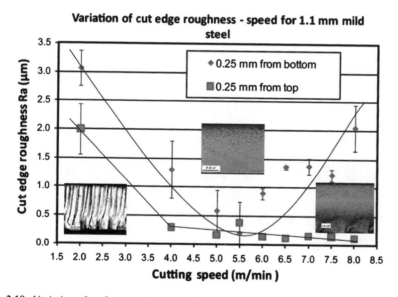

Fig. 2.19 Variation of surface roughness at 0.25 mm from top and bottom surfaces with cutting speed during fiber laser cutting mild steel with oxygen jet. Insert photographs were taken from the lower regions at cutting speeds of 2.0, 5.5 and 7.0 m/min, respectively (reprinted from Powell et al. [56] with permission. Copyright © Elsevier)

because of excessive oxidation and decreases with increase in cutting speed in reactive laser cutting [26, 67] and laser fusion cutting [45]. An optimum cutting speed exists for minimum surface roughness (striation-free) for reactive laser cutting of thin workpiece [56, 60, 78, 79] as shown in Fig. 2.19.

Change in the wavelength of striation pattern is controversial. Ivarson et al. [24], Chen et al. [67], and Schuöcker [49] reported that the striation wavelength gradually increases with increase in cutting speed at constant laser power when cutting mild steel with oxygen jet. Kaplan et al. [27], Poprawe et al. [43], and Ledenev et al. [80] concluded that the wavelength of periodic striations in zone 1 is independent of the cutting speed, but strongly correlates with the beam radius at the workpiece surface in reactive laser cutting of mild steel [27] and reduces with increase in gas pressure in laser fusion cutting [80]. Hirano et al. [45] found that the striation wavelength gradually decreases with increase in cutting speed during laser fusion cutting.

The formation of striation is the result of dynamic change in cut front thickness and temperature which can be brought by thermal dynamic instability (cyclical heating) or hydrodynamic instability (dynamic melt removal). There are several mechanisms for striation formation proposed as follows:

1. Thermal dynamic instability

The energy input fluctuates in laser cutting. It may come from (1) temporal fluctuations of laser power (either due to periodic absorption, periodic distortion of laser power, and coupling of reflected laser radiation with laser resonator) [23, 49]; (2) cyclical burning in reactive laser cutting due to oxidation dynamics at low speed [24–26, 28, 56, 59, 60]. The temporal fluctuations of laser power can be minimized by stabilizing laser output power and mode [23]. Instability in thermal dynamic results in localized heating and intermittent melting in the kerf front in the region where the kerf front wall becomes nearly vertical [45].

• Cyclical burning in reactive laser cutting

This was first observed by Arata et al. [25]. A sideways burning cycle of ignition–burning–extinction (as shown in Fig. 2.20a) occurred in reactive laser cutting when the cutting speed is below the velocity of the oxidation front, a critical speed at typically 2 m/min (or 33 mm/s) for mild steel. Oxidation front moves at velocity of V_M faster than laser beam velocity of V during burning process (process II), but slower during extinction process (process III).

The oxidation when the cutting speed is below the velocity of the reaction front is a dynamic process. When the oxidation front moves well ahead of laser spot (in extinction process III) due to the release of exothermic heat during burning process II, it increases the instantaneous thickness of FeO and decreases the oxygen gradient as shown in Fig. 2.20b. The reduction in oxygen gradient with growth of liquid volume (thickness) during burning process contributes to the extinction. Reignition occurs after the molten layer is removed from cutting zone and laser beam recatches the cutting zone [24].

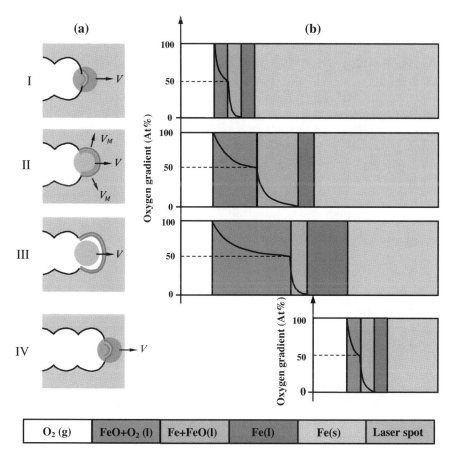

Fig. 2.20 Schematic illustration of (**a**) top view, (**b**) side view and oxygen gradient across melt film during a cycle of (I) ignition, (II) burning, (III) extinction, and (IV) reignition during formation of striation by sideways burning (drawn in reference to [24, 25])

2. Hydrodynamic instability

The ideal flow of the melt out of kerf is laminar flow with uniform thickness and temperature. However, the melt layer oscillation which leads to striation formation at high cutting speed with inert gas could occur due to:

- fluctuation in gas flow (to be discussed in Section "Assist Gas" in details) [49];
- transition of gas flow from laminar to turbulent due to increase in angle of inclination of cut front, which could result in reaction energy intensity instability and metal removal instability [81];
- increase in molten layer thickness, which may lead to spatial distortion on the cutting edge due to liquid oscillation and turbulent flow near the bottom surface [56, 81];

- the onset of ripple formation on the cut edge is caused by the time-dependent movement of the width of the cutting front as a response to fluctuations of the processing parameters [52];
- melt flow transition from one-dimensional to two-dimensional melt flow due to evaporation, which causes liquid layer oscillation [23, 81];
- melt flow is hydrodynamically unstable if the flow is dominantly driven by pressure gradient of the gas flow [53];
- cyclical formation, growth of liquid droplet, which is separated (broken away) from the melt layer when the weight of the forming droplet and the melt flow's starvation pressure overcome the surface tension [82] or the critical radius of droplet is reached [80], the critical drop size is reached when the balance between capillary forces and forces exerted on the melt by the gas stream in the vicinity of the upper edge of the cutting front is achieved [83];
- the periodic growth of the humps (as shown in Fig. 2.10), which start to flow when the gas force exceeds the surface tension force [45];

Because the cut surface quality is important in some applications, the following strategies have been developed to reduce or eliminate the striation formation:

- High-pressure cutting, where the high gas pressure efficiently ejects the melt film, resulting in a thin melt film. Thereby, the amplitudes of the melt film oscillations are limited [23].
- Low pressure cutting, applying subsonic gas jets, where the melt film thickness is relatively large, but the smooth gas flow reduces the melt film oscillations [23].
- The melt flow dominantly driven by shear force is stable and can be achieved by either decreasing cutting speed or increasing gas velocity. This is proven only for the striation formation at high cutting speed [53].
- Cut at speed higher than the speed of combustion-front motion [25, 72].
- Carefully control the pulse frequency of laser beam (in the order of twice the striation frequency under the CW laser cutting condition) [84, 85].
- Material is mainly removed by vaporization due to the suppression of melt flow formation, which can be achieved with high brightness and the small beam divergence of the fiber laser [28, 60, 79] or pulsed laser with high average power and high pulse repetition rate [86].

Dross

When the momentum of molten layer transferred from the pressurized gas is higher than the surface tension force, clear cut is made by removing the molten metal from the bottom of kerf [87]. Formation of dross at the bottom cut edge is the result of high surface tension and viscosity of the melt film. Therefore, the melt cannot be expulsed completely from the bottom of the kerf.

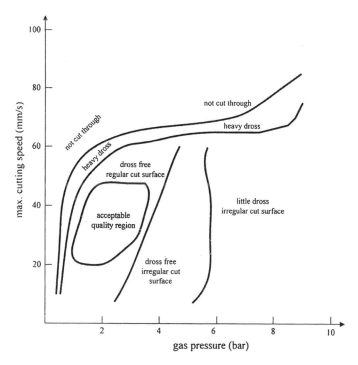

Fig. 2.21 Process window for high-quality cut in reactive laser cutting mild steel with thickness of 3 mm (reprinted from Chen [90] with permission. Copyright© Elsevier)

Similar to striation formation with respect to cutting speed in reactive laser cutting, there is an optimal cutting speed to achieve minimum dross cut as shown in Fig. 2.21. The larger amount of dross produced at cutting speed below or above this optimal cutting speed is due to the solidified large amount of backward flow of melt and strong disturbance of the melt flow [88], respectively.

In order to deform, break up, or detach the melt droplet at lower cut edge by shear action of the gas to achieve a dross-free cut, the Weber number (We) of the melt in the lower part of the cutting front must exceed a critical value [46].

Dross formation when cutting thick workpiece when laser beam is focused at the top surface is due to insufficient energy. This leads to difficulty in separating melt from the lower cut edges. Increase in laser power can eliminate the dross at the expense of poor cut quality because of poor beam-mode quality of CO_2 laser at high laser power [89].

The dross attachment on the lower cut edge can be reduced by increasing the depth of flow separation as discussed in Section "Separation of Boundary Layer Flow".

Melt flow velocity increases and the melt film thickness decreases with increase in assist gas pressure or kerf width. This leads the boundary layer separation point moving down the cut kerf. Therefore, the dross attachment on the lower cut edge reduces with increase in assist gas pressure and cut kerf width in laser fusion cutting [54, 90].

During reactive laser cutting of thick workpiece, the oxygen jet is kept at low pressure (maintaining the gas flow developed in the cut channel to be subsonic, transonic, or slightly supersonic) in order to control the regime of metal burning. The ambient air is entrained in the oxygen flow [73, 91], which increases the impurity of oxygen jet and retardation of oxidation reactions.

Vortex is formed when pressure in the peripheral nozzle is lower than a critical value. The vortex flow can entrain and accumulate the liquid melt, which restrains its removal from the cut even induces the reverse motion of the melt and is responsible for slagging at the bottom of the cutting kerf and adhesion of dross [73, 75].

Heat-Affected Zone (HAZ)

Laser beam cutting is a thermal process. The heat conduction loss to the surrounding solid part results in temperature rise in the subsurface at a depth from the cut surface [32], which may results in change in the microstructure and phase composition such as martensitic transformation, change in grain size, and carbide formation in the HAZ. These changes are normally undesirable because of higher tendency of cracking, surface hardening, decrease in weldability, corrosion resistance, and fatigue life [87, 92, 93].

The width of HAZ increases with increase in workpiece thickness because of the higher heat conduction loss when cutting thicker workpiece [94].

Increase in cutting speeds causes the energy density to decrease and less heat conducted to the bulk workpiece resulting in reduction in thickness of HAZ layers if no resolidified material is attached to the cut edge [93].

The width of HAZ is affected by the type of cutting nozzles used. The HAZ obtained when using conical coaxial cutting nozzle is larger than that obtained by using the supersonic cutting nozzle. The reduction in HAZ by using Laval nozzle is more significant when cutting thicker workpiece [94].

The thickness of HAZ layer decreases with the increase in gas pressure during laser fusion cutting due to the effective removal of the melt from the laser-material interaction region by the strengthened shearing stress of the gas flow and the strengthened cooling effect [93].

The melting point of some products of the chemical reaction during reactive laser cutting is very high, for example, 2,017, 2,230, and 2,435 °C for Al_2O_3, AlN and Cr_2O_3 [36, 94], respectively. They are repeatedly fractured due to the turbulent nature of the melt flow out of the cut zone and dispersed in the melt, which results in increase in the viscosity and surface tension of the melt. Therefore, the velocity of melt flow is reduced, which causes the larger HAZ [94].

The dross attached at the lower cut edge is the melt that is not completely extracted by the assist gas from the cut kerf and solidifies on the cut edge. This produces significant HAZ at the lower region of cut edge as shown in Fig. 2.22.

Fig. 2.22 HAZ at the lower region of the cut surface due to heat by the adhered resolidified layer (reprinted from Riveiro et al. [95] with permission. Copyright © Elsevier)

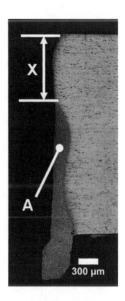

The amount of resolidified material on the cut surface increases with increase in laser power and cutting speed because of the higher rate of melting and higher recoil pressure which expels more molten material backwards toward the cutting front [46]. The proportion of distance X (the distance from the upper surface of the plate up to the beginning of the HAZ) to the thickness of workpiece increases with increase in laser power, gas pressure [95], and optimum pulse frequency [96].

HAZ due to the heat conduction can be reduced when cutting with pulsed laser at high pulse frequency [93, 97] and low duty cycle [97], because of lower heat input.

However, the HAZ due to the adhesion of resolidified melt layer on the cut surface during pulsed laser cutting of ceramics increases with increase in pulse frequency because of the thicker resolidified layer at the end of laser pulse [22].

Fig. 2.23 Ranges of cutting speed and cut quality (reprinted from Poprawe and König [43] with permission. Copyright © Elsevier)

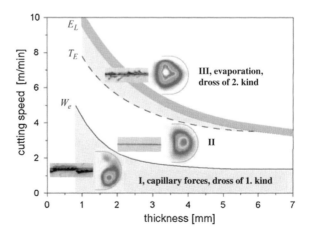

The plasma formation also increases the HAZ, and this is more significant when cutting titanium alloys compared with mild steel because of the significantly lower thermal conductivity of titanium alloys [98].

2.3.2.4 Processing Parameters

Cutting Speed

Laser cutting can be conducted in a wide range of cutting speeds, which strongly depends on the workpiece material and its thickness as shown in Fig. 2.23 [43]. The maximum cutting speed decreases with increase in workpiece thickness. There are three regions in Fig. 2.23 with respect to cutting speed for laser fusion cutting [43] as follows:

- In Region I at low cutting speed range where Weber number We $\approx$ 1, the capillary forces become comparable with the inertia of the melt. Therefore, the melt flow does not separate completely from the cut kerf and dross attachment occurs at the bottom of cut edge.
- In Region III at high cutting speed range from T_E where vaporization of melt film occurs to the maximum cutting speed limited by energy (E_L) for a given thickness of workpiece, melt flow is interrupted by the onset of vaporization. The melt flow becomes two-dimensional due to the vapor pressure (as shown in Fig. 2.16) leading to rough cut at the lower region of the cut surface [66].
- In Region II at the middle cutting speed range, workpiece is melted and melt flow is stable one-dimensional.

The maximum cutting speed is limited by the workpiece thickness and laser power [43]. At a given laser power, the maximum cutting speed decreases with workpiece thickness. It should be noted that the evaporation contribution to the maximum cutting speed reduces with increase in workpiece thickness [48].

The optimal cutting speed is defined as the highest speed for the best quality cut. It is generally 80–90 % [99] or between 70 and 80 % [78] of the absolute maximum cutting speed.

The maximum cutting speed without evaporation can be theoretically calculated by combining Eqs. (2.20), (2.21), and (2.22) as follows [66]:

$$V_{\text{Melt}} = \sqrt{\frac{(T_E - T_M)K_M\sqrt{P_g}}{((T_M - T_0)C_p + L_M)\sqrt{\rho_m^3} \cdot d \cdot \beta_B}} \qquad (2.27)$$

The empirical relationship between incident laser power (P), cutting speed (V), thickness of workpiece (d), and width of kerf or diameter of laser spot (w_k) can be written as follows:

$$P = B \cdot w_k^{0.01} d^{0.21} V^{0.16} \qquad (2.28)$$

which is valid to a high degree of accuracy for laser cutting of different workpiece materials (steel, glass, plastic, wood PVC, etc.) with variation of B only [100, 101].

Laser Beam

Cutting performance is strongly affected by the characteristics of laser beam, which can be operated with different wavelengths, polarizations, and modes, focused with different optics and different focus position relative to workpiece surface.

Polarization of Beam

Due to the difference in absorption with the polarization and incident angle for linearly polarized laser (such as CO_2 laser in Fig. 2.4), the maximum cutting speed when the cutting direction is parallel to the plane of polarization (V_p) is higher than the maximum cutting speed when the cutting direction is perpendicular to the plane of polarization (V_s). Therefore, the plane of polarization should be parallel to the cutting direction for straight-line cut, but a circularly polarized beam should be used for profile or contour cutting in order to avoid the directional effect [6, 102].

Since the p-polarized laser beam is almost completely absorbed at the Brewster's angle (about 87.3° as shown in Fig. 2.4) and change in angle of inclination with cutting speed (as shown in Fig. 2.9), the fastest cutting speed can be achieved with inclination angle of cut front at 2.7° [32].

Because the difference in absorptivity between p-polarized beam and s-polarized beam changes with the workpiece thickness as the change in angle of incidence with workpiece thickness, the ration of maximum speeds, V_s/V_p, decreases with increase in workpiece thickness, reaches a minimum at a critical thickness,

Fig. 2.24 Effect of workpiece thickness on the maximum cutting speed ratio with respect to beam polarization (reprinted from Lepore et al. [103] with permission. Copyright © Elsevier)

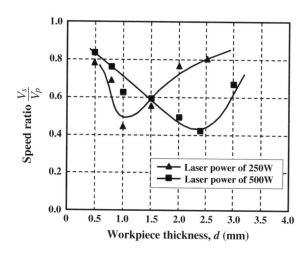

and then increases with increase in workpiece thickness (as shown in Fig. 2.24). The maximum effect of beam polarization is achieved at the critical workpiece thickness. This could be due to the fact that the contribution of exothermic heat is stronger when the workpiece thickness is larger than its critical value during reactive laser cutting [103] or the angle of incidence is equal to Brewster's angle at the critical workpiece thickness.

Muys et al. [104] conducted simulation to verify a model proposed by Niziev and Nesterov [105] and conclude that radial polarization of laser beam can either increase cutting speed for the same thickness or increase the cutting depth for the same laser power.

Wavelength of Laser Beam

It has been reported that the cutting with fiber laser is more effective because of the smaller severance energy (linear energy per unit sheet thickness) during fusion cutting with fiber laser compared with CO_2 laser [21]. Cutting speed is remarkably increased with high-brightness fiber laser compared with CO_2 laser in cutting thin workpiece [21, 106, 107]. The cut quality with fiber laser is worse when cutting workpiece thicker than 5 mm as shown in Fig. 2.25.

However, the cut quality and cutting performance during reactive laser cutting of steel sheet with thickness up to 20 mm are not significantly affected by the wavelength [107], which indicates the contribution of oxidation in reactive laser cutting is more pronounced than the contribution of the characteristics of laser.

The difference between fiber and CO_2 lasers is their wavelength and absorption behavior. The shorter wavelength of fiber laser enables it to be focused for longer distance (longer depth of focus as calculated by Eq. (2.6)) which provides high energy intensity to cause the workpiece vaporization and create a keyhole [108].

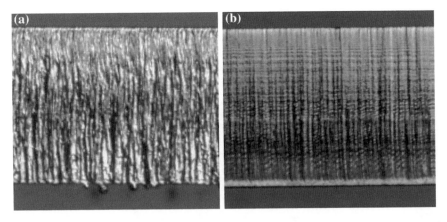

Fig. 2.25 Cut surface of 4-mm stainless steel with laser fusion cutting by (**a**) fiber laser and (**b**) CO_2 laser at power of 2.5 and 3.0 kW, respectively (reprinted from Poprawe et al. [107] with permission. Copyright © Elsevier)

Almost complete melt ejection can be achieved in thin sheet by the vapor pressure gradient, which leads to clear cut [107]. The smaller Brewster's angle with fiber laser (79.6°) compared with CO_2 laser (87.3°) as shown in Fig. 2.5 enables the better laser absorption at larger angle of inclination of cut front (with higher cutting speed), which can be achieved with thin workpiece sheet [31]. In order to achieve the better efficiency in fiber laser cutting thick sheet, the inclination angle of cut front needs to be increased, which can be achieved by adapting the beam oscillation strategy with a longitudinal beam deflection during the cutting process [31].

The more efficient cutting and higher cutting speed with fiber laser is attributed to the better beam coupling and increased laser beam absorption as result of multiple reflections [109, 110], which is more severe with thicker workpiece. The multiple reflection contribution typically starts at the depth of 1–2 mm below the surface and causes the curved and kinked cut front as shown in Fig. 2.9 [56]. It destabilizes the lower cutting zone and leads to coarser striations [109].

Because of the narrower kerf with fiber laser due to its longer depth of focus [31], the gas pressure reduced significantly down through the cut kerf, which lowers the melt flow out of the kerf [111]. Multiple laser beams were developed by Olsen et al. [108] when cutting with fiber laser. The laser intensity is tailored as a central high-intensity melt beam followed by several beams in an appropriate beam pattern, which guide the melt flow out of the cut kerf by creating local evaporation as in the keyhole. Cut quality is improved over a wide range of cutting speeds, and dross-free cut is achieved with this beam configuration compared with single beam cutting.

Pulsed Laser Beam

Pulsed laser provides a high instantaneous power for a period of pulse duration and followed with a period of power off. This offers some benefits in laser materials processing compared with CW lasers [5]:

1. The high pulse peak power results in improved laser coupling for some materials with high reflectivity such as aluminum alloy.
2. The temporal limitation in energy input leads to reduced heating of workpiece and small depth of heat conduction into the workpiece. This is more prominent for materials with low thermal conductivity such as titanium and nickel alloys.

The small heat input with pulsed laser cutting also benefits the reduction in stress-induced cracking in the brittle materials such as ceramics.

These advantages are achieved at the cost of reduced cutting speed at low pulse frequency compared with CW laser cutting [5, 112].

Under the optimized conditions (pulse frequency range of 400–600 Hz and power on-time/off-time ratios of between 5 and 9 to 1), the optimal cutting speed with pulsed laser is higher than that with CW laser by 10 % or more at the same average laser power due to the higher temperature in the cutting zone associated with higher peak laser power in reactive laser cutting mild steel. Furthermore, fluctuation in maximum cutting speed induced by oxygen shock wave during reactive laser cutting is smaller when using pulsed laser compared with CW laser [113].

Laser spot overlap is important for pulsed laser cutting. The spot overlap required for cut decreases with increase in pulse energy and a minimum of 45 % spot overlap is required for clearly through cutting. The kerf width increases and surface roughness decreases with increase in spot overlap [114].

For a given pulse energy, it was found that the maximum cutting speed increases with reduction in pulse duration for cutting nickel-based superalloy [114] as the higher peak laser power with shorter pulse duration. Bicleanu et al. [115] and Lee et al. [116] reported that cutting speed increases with increase in pulse width in pulsed Nd:YAG laser cutting of mild steel and stainless steel.

During pulse mode laser cutting, melt ejection occurs in each laser pulse on-time while melt front cools during pulse off-time. Minimizing dross formation can be achieved in pulsed mode cutting by using low duty cycle pulses which not only provided low heat input that reduced the quantity of melt formed but also imparted sufficient cooling between two successive laser pulses which prevented over-heating of the cut front [97].

Cut quality is reported to be improved by using pulsed lasers [43, 84, 85]. The laser pulse duration (t_{on}) has to be matched to the typical time ($d/_{\bar{v}_m}$) for the propagation of a melt perturbation along the whole cut front down the workpiece thickness in order to allow the melt produced in one laser pulse to be ejected

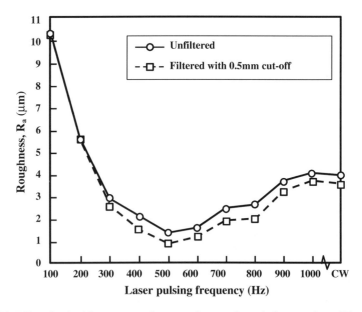

Fig. 2.26 Effect of pulsed frequency on the cut surface roughness in laser cutting mild steel with oxygen (reprinted from King and Powell [84] with permission. Copyright © Elsevier)

completely for dross-free cut. The cycle time $t_{on} + t_{off}$ has to meet the time-averaged energy balance [43].

The interference of perturbation dynamic induced by pulsed laser with the melt flow dynamics enable to minimize the striation formation. The reduction to approximately one-quarter of surface roughness obtained under CW cutting is achieved at the pulse frequency between 400 and 700 Hz (as shown in Fig. 2.26). Higher roughness in the pulsed ranges below and above this range is attributed to the small laser spot overlap and diminished effect of individual laser pulses, respectively. The surface roughness is not significantly influenced by pulse length (mark-space ratio) at the optimum frequency range [84].

The difference between the unfiltered and filtered roughness curves in Fig. 2.26 shows that contribution of a low-frequency periodic roughness (excluded with 0.5-mm cutoff by a phase-corrected high-pass filter) with respect to laser pulsing frequency.

Recrystallization in the heat-affected zone leads to deterioration of surface characteristics. The width of recrystallization zone can be reduced with longer pulsed width at a given pulse energy due to the lower peak power with longer pulsed width [117].

During the pulsed laser cutting of ceramics, the melt is predominantly removed by the momentum transfer from the pressurized gas jet at high pulse frequency and recoil pressure at low pulse frequency [22].

Focusing Laser Beam

Kerf width increases if the focus of the laser spot shifts from the workpiece surface. Minimum kerf width is achieved when the laser spot focus is on the workpiece surface [111]. Wider kerf width can be obtained when laser beam focuses either above the workpiece top surface or close to the bottom surface of the workpiece. This enhances the melt flow velocity in the cut kerf so that the melt flow clears the bottom cut edge before flow separation occurs and with minimal dross attachment on the lower cut edge when cutting thick workpiece provided the energy intensity is sufficient [54].

Due to the relative wider kerf width during laser reactive cutting as the result of sideways burning and low viscosity of melt with FeO, laser beam is recommended to focus at top with oxygen-assisted laser cutting [89].

For cutting thin metal sheet, the laser focus is recommended to be on the top surface of the workpiece to get narrow kerf. However, the melt in narrow kerf is difficult to be removed when cutting thicker workpiece. Therefore, it is suggested that laser beam is focused at bottom of thick stainless steel with nitrogen as assist gas to avoid burr formation [43, 89], with a disadvantage of increased roughness at the lower cut edge [43].

With a single transmitting or reflecting optical element, a laser beam can be focused at two fixed focal points at two different positions, spaced from each other in the direction of material thickness. The concentric circular center surface areas of the optics create a focal point near the bottom surface of workpiece. The amount of laser power focused near the bottom surface of workpiece depends on the diameter of this area. The outer ring area of the optics focuses the remained laser power close to the top surface of the workpiece. The possible thickness capable of being cut with a certain laser power level increases with application of dual focus lens [89].

The cutting speed increases by 23 % and kerf width reduces by 30 % with more perpendicular edges and less dross by using dual focus lens in cut medium section stainless steel with nitrogen compared by using single focus lens. This is attributed to the fact that a long depth of partially focused collimated beam at above and below the upper focal position. This deep-collimated beam increases temperature of the melt at the central sloped cutting zone high enough (below boiling point) to melt the material in the area is not directed irradiated by laser beam, which leads to increase in melt viscosity [118].

Longer focal length is found to reduce the maximum cutting speed by a factor ranging from 1.2 to 2.5 when cutting aluminum alloys [95]. This result can be attributed to a reduction in the power density on the cutting front as the result that spot size increases with increase in focal length.

Assist Gas

The functions of assist gas during laser cutting are to [119]:

1. protect the workpiece from undesirable reactions with the ambient gas and cool the hot cut zone by the forced convection;
2. remove the molten material from cutting front through the narrow kerf to keep the cutting process going. The flow need to be directed with a major component of its velocity in the direction of the expulsion;
3. provide exothermic heating by chemical reaction if reactive gas used.

Table 2.4 Chemical reactivity of common used gases with the common metals

	O_2	N_2
Fe	$Fe + 1/2O_2 = FeO + 275$ kJ/mol (at 2,000 K) [37] $4Fe + 3O_2 = 2Fe_2O_3 + 822.2$ kJ/mol[a][156] $3Fe + 2O_2 = Fe_3O_4 + 1,117$ kJ/mol[a][155]	
Al	$4Al + 3O_2 = 2Al_2O_3 + 1,670$ kJ/mol[a] [156]	$Al + N=AlN + 1050.4$ kJ/mol (at 1,103 K above) [157]
Ti	$Ti + O_2 = TiO_2 + 912$ kJ/mol[a] [156]	$2Ti + N2 = 2TiN$
Cr	$2Cr + 3/2O_2 = Cr_2O_3 + 1163.67$ kJ/mol (at 2,000 K) [37]	
Ni	$Ni + 1/2O_2 = NiO + 244$ kJ/mol[a] [156]	
Cu	$2Cu + 1/2O_2 = Cu_2O + 166.7$ kJ/mol[a] [156] $Cu + 1.2O_2 = CuO + 155$ kJ/mol[a] [156]	

[a] at 293 K for comparison

Table 2.5 Comparison of properties of inert gases for laser fusion cutting (reprinted from Rao et al. [97] with permission. Copyright © Elsevier)

	Ar	He	N_2
Stagnant gas pressure (MPa)	0.6	0.6	0.6
Gas density (kg/m^3)	1.3038	0.1306	0.9776
Specific heat capacity (J/kg K)	520	5,200	1,183
Thermal conductivity (W/m K)	0.0452	0.3765	0.0738
Viscosity (μPa s)	58	48.3	44
Critical gas velocity (m/s)	280	884	322
Shear stress (N/m^2)	850	1,390	800
Heat transfer coefficient (W/m^2 K)	2,080	10,680	3,690

The cooling of the melt due to the gas jet is shown to be negligible compared to the incident laser beam power [50].

These functions are heavily dependent on the nature, pressure, and purity of the assist gas.

Gas Nature

The chemical reactivity of the commonly used gases for laser cutting is listed in Table 2.4. In addition to oxygen, nitrogen is chemically reactive with aluminum and titanium.

The properties of inert gases for laser fusion cutting are listed in Table 2.5. Helium shows the highest values of critical gas velocity, shear stress, and convective heat transfer coefficients.

Fig. 2.27 Effect of oxygen purity on cutting speed during cutting 2.0-mm-thick mild steel at CO_2 laser power of 800 W (drawn in reference to [99])

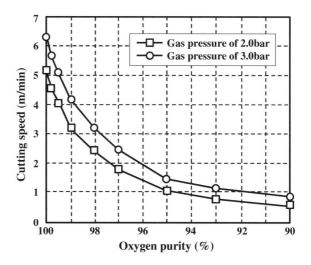

Gas Purity

Contamination of the inert gas only causes degradation of the cut surface, such as dross size, oxidation, or nitriding of the cut face. Cutting mild steel with mixture of two inert gases does not show significant difference with pure inert gas in cutting speed and dross formation at low gas pressure [41]. However, the impurity in oxygen significantly affects the cutting process.

The cutting performance strongly depends on the percentage of impurity in oxygen gas jet but not on the type of impurity gases, no dross-free cut can be obtained at oxygen level lower than 75 % in cutting mild steel of thickness of 3 mm with laser power of 1.5 kW and gas pressure of 2 bar [41].

The cutting speed is very sensitive to the oxygen purity at high purity level (99.998–95 %), 2 % of impurity is found to reduce the optimal cutting speed by more than 50 %. Cutting speed is not significantly affected by further decrease in oxygen purity from 95 to 90 % (as shown in Fig. 2.26) because the heat contribution by oxidation reduces with reduction in oxygen purity [99].

The oxidation is significantly reduced by the formation of a boundary layer at the gas–liquid interface in which the impurity gas is accumulated. This layer acts as a barrier to inhibit the oxidation reaction by lowering the oxygen diffusion rate in this layer [26, 99]. The thickness of the boundary layer can be reduced by increasing the gas pressure. Therefore, the cutting speed is increased by increase in gas pressure at the same oxygen purity level as shown in Fig. 2.27.

Reduction in oxygen purity also significantly reduce the critical cutting velocity above which oxidation is stable without the cycle of ignition-burning extinction [59].

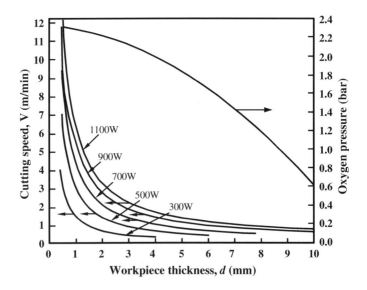

Fig. 2.28 Dependence of cutting speed and optimum oxygen pressure on workpiece thickness for CO_2 laser cutting steel (reprinted from Schuöcker [4] with permission. Copyright © Taylor & Francis)

Gas Pressure

The gas pressure does not show a significant influence on the maximum cutting speed but does have a significant effect on the quality of the cut edge in the laser fusion cutting process [33].

Increase in gas pressure with workpiece thickness (minimum of 8 bar for cutting stainless steel with nitrogen) is required to effectively remove the melt from kerf when cutting with inert gas. However, the gas pressure should be reduced (maximum of 6 bar when cutting mild steel with oxygen) with increase in workpiece thickness when cutting with reactive gas in order to control the cut surface quality as shown in Fig. 2.28 [4, 68, 120].

There are two optimum oxygen pressure ranges to obtain the good cut quality for both CW and pulsed CO_2 laser cutting of mild steel. The upper and lower limits of these two pressure ranges when cutting thicker workpiece are lower than these when cutting the thinner workpiece. Hence, the pressure of the assist oxygen jet is reduced with increase in workpiece thickness at a constant laser power as shown in Fig. 2.28. The reduction in oxygen pressure is much lower than the reduction in cutting speed with increase in workpiece thickness [113].

The optimal cutting speed dramatically increases with increase in oxygen pressure at workpiece up to 2 bar and barely changes with oxygen pressure at workpiece increasing from 3.5 bar for cutting 8-mm-thick steel with both CW and pulsed lasers [113]. Small kerf width is achieved because both the thermal energy and mechanical energy provided by oxygen at this pressure range are sufficient and the melt flow is not macroscopically turbulent but periodic in the lower optimum pressure range and the controlled oxidation due to the reduction in macroscopic turbulence of the thinner molten layer by the increasing gas pressure in the higher optimum pressure range, respectively [113].

Gas Nozzle

Nozzle Types

• Subsonic nozzle

Nozzles which deliver the assist gas are generally subsonic nozzle with converging cross section as shown in Fig. 2.29. The characteristics of gas flow depend on the stagnation pressure, P_0 [121, 122]:

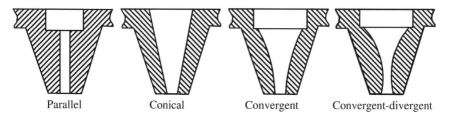

| Parallel | Conical | Convergent | Convergent-divergent |

Fig. 2.29 Cross-sectional geometries of commonly used subsonic coaxial nozzle (reprinted from Fieret et al. [150] with permission. Copyright © SPIE)

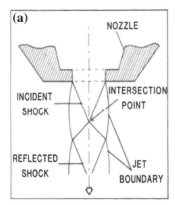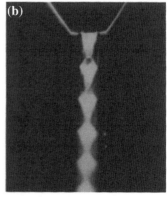

Fig. 2.30 a Illustration and **b** Schlieren image of Prandtl–Meyer wave (reprinted from Fieret et al. [150] with permission. Copyright © SPIE)

Fig. 2.31 Illustration of gas–melt interaction of supersonic gas flow with subsonic nozzle (reprinted from Riveiro et al. [95] with permission. Copyright © Elsevier)

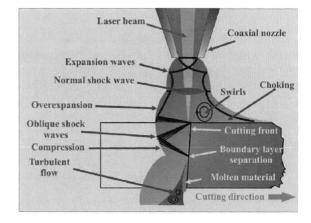

1. $P_0 < 1.89 \, \text{kg/cm}^2 (0.185 \, \text{MPa})$: subsonic flow in which both gas velocity and flow increase with stagnation pressure;

2. $P_0 = 1.89 \, \text{kg/cm}^2 (0.185 \, \text{MPa})$: transonic flow in which gas velocity reaches the maximum value;

3. $P_0 > 1.89 \, \text{kg/cm}^2 (0.185 \, \text{MPa})$: supersonic flow in which gas velocity maintains the sonic velocity while gas flow increases with stagnation pressure. Prandtl–Meyer wave as shown in Fig. 2.30 is formed.

4. $P_0 > 5 \, \text{kg/cm}^2 (0.49 \, \text{MPa})$: supersonic flow in which strong Mach shock disk is formed (as shown in Fig. 2.31). The impingement of the oblique shock wave over the boundary layer produces a pressure gradient that may separate the boundary layer. The gas flow below the detachment point becomes turbulent with eddies and slipstreams that reduce the momentum transfer and the capability of the gas jet to remove the molten material and leads to the poor cut quality (increased dross attachment, heat-affected zone, and rough cut surface

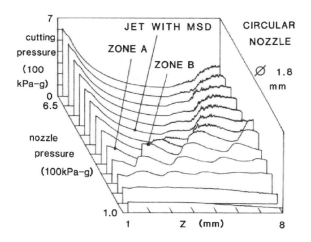

Fig. 2.32 Gas pressure at workpiece as a function of gas pressure in nozzle and distance from the nozzle exit, z for a circular nozzle (reprinted from Fieret et al. [151] with permission. Copyright © SPIE)

as shown in Fig. 2.17) [75, 121, 123]. The gradients of gas density that is produced by non-uniform jet flow results in a change of the refractive index in the gas field (acts like an optics), which leads to the secondary focusing or diverging of the laser beam. The result of interference can affect obviously the melting efficiency and change the mode of the laser beam, which causes poor cutting quality and reduces cutting speed [121].

The cutting pressure as a function of pressure in nozzle and distance between nozzle and workpiece surface (stand-off distance) for a conical subsonic nozzle is shown in Fig. 2.32. The gas pressure drops sharply in a very short distance from the exit of nozzle (with high nozzle pressure). Hence, the stand-off distance should be kept very small (smaller than the diameter of the nozzle), usually 0.3–1.3 mm for a subsonic nozzle in order to maintain the gas pressure [124].

• Supersonic nozzle:

Laval nozzle contains stable, convergent, throat, and divergent sections (Fig. 2.33a). The gas flow is supersonic at the Laval nozzle exit, and the free gas jet stays homogeneous with weak oblique shocks and parallel for a long distance (i.e., long supersonic length as shown in Fig. 2.33b) [122, 125]. Therefore, the toleration of stand-off distance with using Laval nozzle is greater compared with using subsonic nozzle [121].

The higher gas speed exits from Laval nozzle significantly improves the melt flow ejection, which leads to better cut quality in terms of small HAZ and less dross attachment.

Nozzle Integration

Gas nozzle can be integrated coaxially or off-axially with laser beam, or combined as shown in Fig. 2.34. The coaxial nozzles are generally used and are particularly

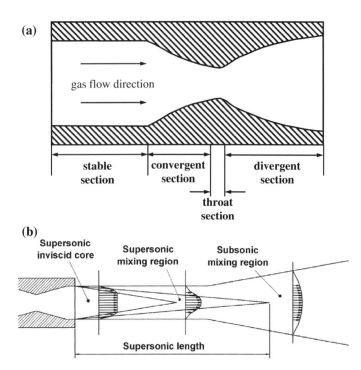

Fig. 2.33 **a** Typical cross section of a Laval nozzle and **b** supersonic length of gas jet by Laval nozzle (reprinted from Quintero et al. [125] with permission. Copyright © Elsevier)

suited for robotic systems performing two- and three-dimensional cutting, while the off-axis nozzle is useful for one-dimensional cutting [119].

The coaxial nozzle is discussed through this chapter. This section is focused on the special features of a single off-axis nozzle and an off-axial nozzle in tandem with the coaxial nozzle.

1. Off-axis nozzle

There are more parameters for adjusting an off-axis nozzle than a coaxial nozzle, including impinging angle between the axes of the laser beam and the gas jet, gap between the nozzle and the workpiece and distance between the impinging point of gas jet and the laser spot on workpiece surface (marked by X in Fig. 2.34b).

The off-axis is mounted after the laser beam in the cutting direction. Maximum cutting speed reaches the peak (50 % higher than that with the coaxial nozzle) with the impinging angle in the range of 35–40° without noticeable degradation in cut edge quality. This improvement is the result that the flow separation due to the shock wave/boundary layer interaction is alleviated [126].

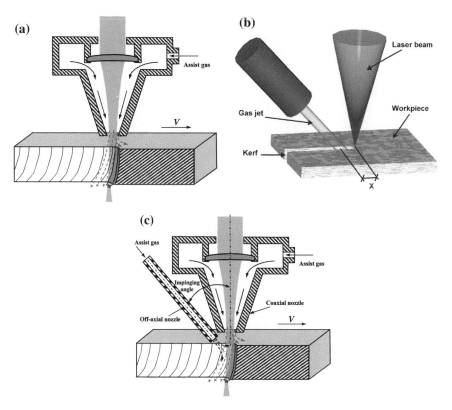

Fig. 2.34 Laser cutting with **a** coaxial, **b** off-axis (reprinted from Quintero et al. [125] with permission. Copyright © Elsevier), and **c** tandem nozzles

When the assist gas jet delivered by an off-axis nozzle impacts onto the cutting front close to 90°, the melt film can be confined near the cutting front in order to have a melt with large Weber numbers and to obtain dross-free cuts [46].

The gap between the nozzle and the workpiece should be smaller than the supersonic length. When a Laval nozzle is used, the impinging angle which determines the aerodynamic behavior of the gas flow inside the kerf should be adjusted in accordance with the Mach number of the nozzle to avoid the formation of a normal and detached shock wave. The distance between the impinging point of the gas jet and the laser beam should be adjusted in such way that a laminar flow is achieved along the whole depth of the cutting front in order to eliminate the formation of a recast layer on the cut edge in laser fusion cutting [123, 125].

2. Tandem nozzles

Use of an off-axis nozzle in tandem with the coaxial nozzle improves the cutting depth up to 20 %. Higher cutting depth is achieved with higher gas pressure, small nozzle-workpiece distance, and impinging angle [127].

Table 2.6 Summary of process parameters affecting cut surface quality (reprinted from Dubey and Yadava [20] with permission. Copyright © Elsevier)

Quality characteristics	Significant factors	Variation of factors to keep minimum value of quality characteristics
Heat-affected zone	Beam energy	Low
	Feed rate	High
	Pulse duration	High
	Pulse frequency	Moderate
	Gas pressure	More
	Workpiece thickness	Low
Taper	Beam energy	Low
	Feed rate	High
	Pulse frequency	Low
	Pulse duration	High
	Workpiece thickness	More
	Focus position	Above the workpiece surface
Surface roughness	Beam energy	Moderate
	Feed rate	Moderate
	Pulse frequency	Moderate
	Gas type	Inert
	Gas pressure	Moderate
Recast layer	Beam energy	High
	Pulse duration	Low
	Gas pressure	High
	Workpiece thickness	Low
	Focus position	Above the workpiece surface
Dross adherence	Gas type	Inert
	Gas pressure	High
	Beam energy	High
	Feed rate	High
	Pulse frequency	Low
Microcracks	Beam energy	High
	Pulse duration	Low
	Gas pressure	High
	Gas type	Inert

With a constant distance between coaxial and off-axis nozzles, material removal rate increases with decrease in impinging angle of the off-axis nozzle. Surface roughness becomes greater at small impinging angles which promotes turbulent flows due to the extended distance from the off-axial nozzle to the erosion-cutting front. The dross height is also affected by the impinging angle of off-axis angle [128].

Higher cutting speed with use of an off-axis nozzle in tandem with the coaxial nozzle in laser cutting is attributed to more effective momentum transfer. The formation of Cr_2O_3 in the melt during cutting stainless steel with oxygen jet is suppressed with the off-axis nozzle [128].

The application of an off-axis nozzle with coaxial cutting head is also reported to prevent ambient gas entrained in the gas flow and to eliminate the formation of vortex in the melt [73].

The effect of processing parameters on the cut quality is summarized in Table 2.6 for cutting with Nd:YAG laser [20].

2.3.3 Workpiece Aspects for Laser Beam Machining

2.3.3.1 Workpiece Thickness

There is a clear maximum material thickness for a particular laser–material combination for both laser fusion cutting and oxygen-assisted laser cutting, beyond which the cutting mechanism breaks down and cannot be reestablished at any speed because the relative increase in thermal losses from the cut zone as the cutting speed is decreased [64, 129]. The maximum thickness of workpiece is limited by the laser power and the shape of the focused laser beam as shown in Fig. 2.27 [4, 43]. It is recommended to increase nozzle diameter with increase in workpiece thickness in order to increase the kerf width [36, 68].

The maximum thickness of a workpiece can be cut at a given laser power decreases with decrease in waveguide attenuation as a fact that fraction of laser radiation transmitted through the kerf without absorbed by the cut front increases with decrease in waveguide attenuation [48].

There are some strategies developed to increase the maximum thickness that can be cut with the available laser power.

Dual Laser Beams

Laser beam is separated by a small distance trailing each other. The first beam partially penetrates the moving workpiece and forms a blind cylindrical keyhole. The second beam impinges the molten region produced by the first beam and further heats it vaporizing some material and superheating the reminder as shown in Fig. 2.35a. Improvement is achieved in terms of higher cutting speed and larger workpiece thickness as shown in Fig. 2.35b, which is attributed to improvement in beam absorption and modification of conduction characteristics.

Spinning Laser Beam

The laser beam is spun by a rotating window which is placed in the beam path at an angle to the beam axis, which causes the laser beam to form a spiral path when it travels across the workpiece. A linear relationship exists between the workpiece thickness and spin speed [130]. It has been demonstrated that mild steel and stainless steel with thickness up to 25 mm can be cut with spinning beam with 1.8 kW CO_2 laser.

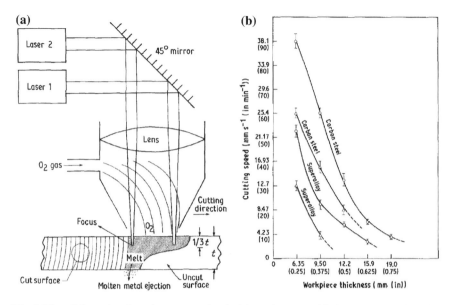

Fig. 2.35 **a** Schematic of cutting process by dual laser beams and **b** improvement of cutting speed compared with a single beam (reprinted from Molian [151] with permission. Copyright © Springer)

Laser-Assisted Oxygen Cutting (LASOX)

Reactive laser cutting of thicker steel plates requires higher-power laser, which not only increases capital cost but also often results in poor cut surface quality in terms of large heat-affected zone and severe striation. In order to generate extra heat, a process named LASOX was developed by O'Neill et al. at the University of Liverpool [131].

Differentiated from the conventional oxygen-assisted laser cutting, up to 80 % of total cutting energy is contributed by the exothermic heat generated from the combustion of iron. Laser beam is used to preheat the surface of workpiece to the ignition point ($\sim$1,237 K). In order to completely use exothermic energy, the oxygen jet footprint is slightly smaller than the laser beam footprint as shown in Fig. 2.36. Thick mild steel plate of 50 mm can be cut at the incident laser power level of 1 kW.

2.3.3.2 Workpiece Materials

Ferrous Metals

The mild steel is easily cut with the oxygen as assist gas because of the contribution of exothermic heat. The oxide formed is easily removed from the kerf due

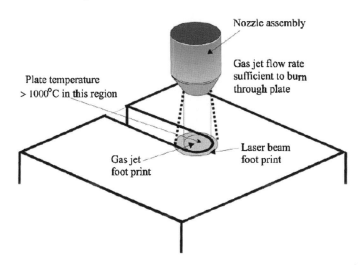

Fig. 2.36 Schematic illustration of laser beam/oxygen footprint configuration in LASOX process (reprinted from O'Neill and Gabzdy [131] with permission. Copyright © Elsevier)

to its low viscosity, and no significant dross is attached to the cut edge. Oxide layer is formed on the cut surface.

In reactive laser cutting of stainless steel, the oxide, Cr_2O_3, formed with a high melting point of 2,435 °C that increases the viscosity and surface tension of the melt. The melt flow velocity is diminished, and the continuous oxidation of the melt cannot be achieved because oxygen cannot diffuse through the Cr_2O_3 layer [36].

The surface of stainless steel cut with oxygen as assist gas is not weldable, and the oxide layer must be removed to eliminate the porosity that occurs in welding [66]. Generally, high-pressure nitrogen is used as assist gas to produce oxide-free and dross-free cuts with laser focused at the bottom surface of workpiece [132].

Non-Ferrous Metals

It is difficult to cut aluminum alloys with laser because of their (1) high reflectivity, (2) high thermal conductivity, and (3) self-extinguishing oxidation reaction with using oxygen as assist gas which is attributed to the fact that aluminum oxide forms a seal on the cut front, which prevents oxygen from penetrating into the melt for further reaction before it is fractured. It is required to increase laser power in CW mode to initiate the cut in the first stage of the process; however, back-reflected beams entering into the laser cavity can damage the laser cavity, cavity optics, or beam delivery optics [95]. Therefore, pulsed laser with high peak power is beneficial for cutting aluminum alloys.

The cutting of aluminum alloy is strongly affected by the assist gas nature as shown in Table 2.7. Oxygen, nitrogen, and compressed air react with aluminum to

Table 2.7 Variation of cutting speed (V) surface roughness (Ra) and HAZ extension in descendent order with assist gas of aluminum alloy (reprinted from Riveiro et al. [94] with permission. Copyright © Elsevier)

	Cutting speed, V	Cut surface roughness, Ra	HAZ extension
(+)	Ar	O_2	O_2
↓	O_2	N_2	Air
(−)	Air	Air	N_2
	N_2	Ar	Ar

produce a large amount of oxides and/or nitrides. These oxides/nitrides attached to the lower cut edge as clinging dross increase the surface roughness and HAZ. Ar is more efficient assist gas to produce the best quality because of the increase in viscous friction with the melt as a result of its highest dynamic viscosity and density [94].

In cutting titanium alloys, a thin layer of hard and brittle oxide or nitride is formed on the cut surface when oxygen or nitrogen is used as assist gas. Thicker HAZ layers are produced due to the heat released from the nitriding and oxidation. Microcracks are formed on the cut surface as the results of tension forces on the laser cut surface and the brittleness of the titanium oxide and titanium nitride. The amount and dimensions of the microcracks can be reduced by increasing the cutting speed. The best corrosion resistance of the cut surface is achieved when cutting with Ar as assist gas [93].

The cut surface of titanium alloy is well protected by using He or Ar as assist gas. The maximum cutting speed is higher and cut edges are straight when cutting with Ar as assist gas compared with cutting with He as assist gas. The wavy cut edges and lower maximum cutting speed when cutting with He as assist gas are attributed to the lower cut front temperature as the result of high convective heat transfer coefficient of He as shown in Table 2.5. The low cut front temperature prevents melt front propagating ahead of the laser spot and limits the melt ejection from the cut zone at higher cutting speeds [97].

Ceramics

Alumina undergoes the following changes when it is subjected to heating [133]:

- Melting at $2,327\,\text{K} \leq T < 3,500\,\text{K}$,
- Dissociation to AlO (g), Al (l), Al(g), and $Al_2O(g)$ at $3,250\,\text{K} < T < 5,000\,\text{K}$,
- Formation of aluminum vapor and atomic oxygen at $T > 5,000\,\text{K}$.

Recoil pressure generated due to the dissociation process provoked expulsion of the melt film formed due to melting between 2,327 and 3,500 K [133]. Oxygen is normally used as the assist gas for cutting alumina to reduce the oxygen depletion and prevent alumina being discolored during laser fusion cutting [134].

One of the main problems in the laser cutting of ceramics is the formation of cracks induced by thermal stresses which are generated in the workpiece due to the local heating of the workpiece during laser irradiation, the brittle nature of ceramics, and the unique properties of ceramic materials (i.e., the high thermal expansion coefficient and low thermal conductivity) [123, 134].

The tendency of cracking when cutting 2-mm-thick alumina with CW CO_2 laser decreases with increase in laser power and decrease in feed rate, which lead to the reduction in temperature gradient [135]. An empirical equation is established for the boundary of cracking in terms of the operating conditions, the material properties, and thickness of the specimen as

$$P = 1.40 \times 10^{23} \rho_w \cdot g^{2.41} d^{2.41} r_L^{2.82} V \cdot c_w^{-1.41} \alpha_w^{1.41} \qquad (2.29)$$

Crack-free cut can be achieved with pulsed laser with short pulse duration for cutting thin ceramic workpiece due to the lower heat input [135]. Cracking is inevitable when the pulse duration increases for cutting thick ceramic workpiece. Multiple passes with short pulse duration can be applied for cutting thick workpiece with low productivity [136–138].

Non-Homogenous Materials

Coated Metallic Materials

A thin layer of zinc and/or aluminum coated on both sides of sheet steels provides good corrosion resistance, but the higher light reflectivity and thermal conductivity, and the lower melting point of the coating layer than that of the substrate materials impose some difficulties and limitations on laser cutting [139].

Cutting speed is reduced when cutting coated steel because of the high reflectivity of the coating in CW laser cutting; however, the effect of coating on cutting speed with pulsed laser is not as significant as with CW laser [112].

Oxygen is quite effective as an assist gas for laser cutting the coated steel sheets as far as the cutting speeds are concerned because of the exothermic reaction between aluminum and oxygen. However, localized overheating and dross attachment are encountered with oxygen-assisted laser cutting. Dross- and oxidation-free cut can be achieved with high-pressure nitrogen [129].

Metal Laminates

Metal laminates are typically aluminum-based, with an aluminum sheet thickness in the order of 0.25 mm. Up to six aluminum sheets may be encountered in one laminate, which are filled up with a synthetic material, such as epoxy embedded glass or aramid fibers, or polypropylene. They can be applied in aeroplanes with advantages of enhanced fatigue resistance and weight reduction [140].

These materials can be cut with laser at the same speed as homogeneous aluminum alloys, with some damage on the synthetic layers and dross attachment occurring. The damage depth of the synthetic layer is significantly reduced with increase in cutting speed up to 6 m/min. Primary interaction between laser beam and synthetic materials is not dominant in the damage extent of the synthetic layers. Heat input into the synthetic layers via the metal layers is most likely to be dominant [140].

Metal Matric Composite Materials (MMCs)

Problems associated with laser cutting composites are the differences between matrix and reinforcements in laser absorption, melting point, latent heat of fusion, and specific heat capacity, which influence melting dynamics at the cut edges during the cutting process [141].

The material removal process in laser cutting SiC fiber–reinforced Al6061 composites (0/90 two-ply laminate) is melting of metal matrix and vaporization of SiC fiber [142].

There is a critical energy density for the cutting of the fiber and matrix, E^f_{crit} and E^m_{crit}, respectively, with $E^f_{crit} > E^m_{crit}$ in the case of SiC fiber–reinforced aluminum alloys. When the laser beam energy is higher than the critical value, E^f_{crit}, both the fiber and the matrix are removed simultaneously; however, when the laser beam energy ranges from E^m_{crit} to E^f_{crit}, only the matrix is removed. Consequently, at first, aluminum alloy matrix is melted and then evaporated and/or blown by the assist gas jet. Then, the bare SiC fiber is directly irradiated by the laser beam and then evaporated. Therefore, it is clear that the laser cutting phenomenon of metal matrix composites is strongly influenced by the laser cutting phenomena of the individual components [142].

In the particulate reinforced metal matrix composites, the specific heat capacity increases and the thermal conductivity decreases with increase in the volume fraction of the reinforcement particles, such as Al_2O_3 and B_4C, these lead to large kerf width. Al_2O_3 and B_4C particles remain in solid phases close to the cut edges. Al_2O_3 particles undergo high temperature reactions forming AlO at cut edges, whereas B_4C particles appear to be round and are present at the cut edges as solid particles [141].

Fiber-Reinforced Plastics (FRP)

Fiber-reinforced plastics, containing reinforcing fiber (glass fiber or carbon fiber etc.) and polymer matrix (GFRP and CFRP), offer high specific strength and are widely used for aerospace and sports industries. The alternative orientation of fiber and matrix creates the difficulties for conventional machining.

Laser beam machining is a suitable technology to cut FRP efficiently with minimum material waste, no tool wear, low overall distortion, or part damage. Nd:YAG and fiber lasers with wavelengths of 1.06 and 1.07 µm, respectively, are transparent to glass and therefore are not suitable for cutting glass fiber-reinforced plastics [143]. CO_2 laser is operated at a wavelength of 10.6 µm that can be effectively absorbed by most organic materials [144], and therefore, is a suitable heat source for cutting FRP.

Because of the different properties of fiber and matrix, the energy required to vaporize the fibers is higher than that required for the matrix. This is more prominent for cutting carbon fiber-reinforced plastics because of the high disassociation temperature and high thermal conductivity of carbon fiber. Large amount of resin matrix is vaporized in the process. This causes delamination and matrix recession of the composite. The different removal rate between fibers and matrix produces the cut surface with fibers devoid of the matrix material [143].

For a given laser power, the maximum cutting speed achievable for CFRP is much lower than that for GFRP. It is visible on the cut surface, and the length of protruding fibers decreases with increase in cutting speed [145].

HAZ decreases with increase in cutting speed and assist gas pressure but independent of type of gas. Higher laser power leads to increase in the HAZ because of higher heat input [143].

Similar to that in cutting metallic workpiece, HAZ is reduced by using pulsed laser beam compared to using CW laser, due to the effective cooling between the pulses [143, 144].

2.4 Concluding Remarks

Laser beam is a unique machining source which can cut materials by photo-thermal processes, in which materials are separated by controlled fracture or locally removed by melt ejection, vaporization or ablation mechanisms.

In the melt and ejection process, a layer of melt (cut front) is formed and extends through the workpiece thickness, kerf is produced after the melt is removed by a pressurized gas jet. This process is able to cut large variety of materials, such as metals, ceramics, composites and glasses, etc.

The cut quality and maximum cutting speed are strongly affected by the (1) characteristics (wavelength, power, focusing and pulsing) of the laser beam, (2) the optical and thermophysical properties of the material, (3) thickness of workpiece and (4) type and pressure of assist gas.

The technology of laser cutting will continue to grow into the future as more powerful, flexible, efficient and compact systems become.

References

1. Meijer J (2004) Laser beam machining (LBM), state of the art and new opportunities. J Mater Process Technol 149:2–17
2. Tunna L, O'Neill W, Khan A, Sutcliffe C (2005) Analysis of laser micro drilled holes through aluminium for micro-manufacturing applications. Opt Lasers Eng 43:937–950

3. Li L (2000) The advances and characteristics of high-power diode laser materials processing. Opt Lasers Eng 34:231–253
4. Schuöcker D (1989) Laser cutting. Mater Manuf Process 4:311–330
5. Olsen FO, Alting L (1995) Pulsed laser materials processing, ND-YAG versus CO_2 lasers. CIRP Ann—Manuf Technol 44:141–145
6. Elijah Kannatey-Asibu J (2009) Principles of laser materials processing. Wiley, Hoboken
7. Yalukova O, Sárady I (2006) Investigation of interaction mechanisms in laser drilling of thermoplastic and thermoset polymers using different wavelengths. Compos Sci Technol 66:1289–1296
8. Bergström D, Powell J, Kaplan AFH (2007) A ray-tracing analysis of the absorption of light by smooth and rough metal surfaces. J Appl Phys 101:13504
9. Petring D, Abels P, Beyer E (1988) Absorption distribution on idealized cutting front geometries and its significance for laser beam cutting. Proc SPIE 1020:123–131
10. Kwon H, Yoh JJ (2012) Polarized reflectance of aluminum and nickel to 532, 355 and 266 nm Nd:YAG laser beams for varying surface finish. Opt Laser Technol 44:1823–1828
11. Mv Allmen (1976) Laser drilling velocity in metals. J Appl Phys 47:5460–5463
12. Chan CL, Mazumder J (1987) One-dimensional steady-state model for damage by vaporization and liquid expulsion due to laser-material interaction. J Appl Phys 62:4579–4586
13. Chang JJ, Warner BE (1996) Laser-plasma interaction during visible-laser ablation of metals. Appl Phys Lett 69:473–475
14. Tam SC, Williams R, Yang LJ, Jana S, Lim LEN, Lau MWS (1990) A review of the laser processing of aircraft components. J Mater Process Technol 23:177–194
15. Segall AE, Cai G, Akarapu R, Romasco A, Li BQ (2005) Fracture control of unsupported ceramics during laser machining using a simultaneous prescore. J Laser Appl 17:57–62
16. Tsai C-H, Chen H-W (2003) Laser cutting of thick ceramic substrates by controlled fracture technique. J Mater Process Technol 136:166–173
17. Kalyanasundaram D, Shrotriya P, Molian P (2010) Fracture mechanics—based analysis for hybrid laser/water jet (LWJ) machining of yttria-partially stabilized zirconia (Y-PSZ). Int J Mach Tool Manuf 50:97–105
18. Barnes C, Shrotriya P, Molian P (2007) Water-assisted laser thermal shock machining of alumina. Int J Mach Tool Manuf 47:1864–1874
19. Tsai CH, Liou CS (2003) Fracture mechanism of laser cutting with controlled fracture. J Manuf Sci Eng, Trans ASME 125:519–528
20. Dubey AK, Yadava V (2008) Experimental study of Nd:YAG laser beam machining—An overview. J Mater Process Technol 195:15–26
21. Mahrle A, Lütke M, Beyer E (2010) Fibre laser cutting: beam absorption characteristics and gas-free remote cutting. Proc Inst Mech Eng C: J Mech Eng Sci 224:1007–1018
22. Quintero F, Varas F, Pou J, Lusquiños F, Boutinguiza M, Soto R, Pérez-Amor M (2005) Theoretical analysis of material removal mechanisms in pulsed laser fusion cutting of ceramics. J Phys D Appl Phys 38:655–666
23. Olsen FO, Alting L (1989) Cutting front formation in laser cutting. CIRP Ann—Manuf Technol 38:215–218
24. Ivarson A, Powell J, Kamalu J, Magnusson C (1994) The oxidation dynamics of laser cutting of mild steel and the generation of striations on the cut edge. J Mater Process Technol 40:359–374
25. Arata Y, Maruo H, Miyamoto I, Takeuchi S (1979) Dynamic behaviour in laser cutting of mild steel. Trans Jpn Weld Res Inst 8:15–26
26. Chen K, Yao YL, Modi V (1999) Numerical simulation of oxidation effects in the laser cutting process. Int J Adv Manuf Technol 15:835–842
27. Kaplan AFH, Wangler O, Schuöcker D (1997) Laser cutting: fundamentals of the periodic striations and their on-line detection. Lasers Eng 6:103–126
28. Sobih M, Crouse PL, Li L (2007) Elimination of striation in laser cutting of mild steel. J Phys D Appl Phys 40:6908–6916

29. Kaplan AFH (1996) An analytical model of metal cutting with a laser beam. J Appl Phys 79:2198–2208
30. Fomin VM, A.G. Malikov, Orishich AM, Shulyat'ev VB (2011) Energy conditions of gas laser cutting of thick steel sheets. J Appl Mech Tech Phys 52:340–346
31. Mahrle A, Beyer E (2009) Theoretical aspects of fibre laser cutting. J Phys D Appl Phys 42:175507
32. Prusa JM, Venkitachalam G, Molian PA (1999) Estimation of heat conduction losses in laser cutting. Int J Mach Tool Manuf 39:431–458
33. Duan J, Man HC, Yue TM (2001) Modelling the laser fusion cutting process: I. Mathematical modelling of the cut kerf geometry for laser fusion cutting of thick metal. J Phys D Appl Phys 34:2127–2134
34. Schuöcker D (1986) Theoretical model of reactive gas-assisted laser cutting including dynamic effects. Proc SPIE 650:210–219
35. Ivarson A (1993) On the physics and chemical thermodynamics of laser cutting. PhD thesis, Lulea University of Technology, Sweden
36. Powell J (1998) CO_2 Laser Cutting. Springer, London
37. Ivarson A, Powell J, Magnusson C (1991) The role of oxidation in laser cutting stainless and mild steel. J Laser Appl 3:41–45
38. Hsu MJ, Molian PA (1994) Thermochemical modelling in CO_2 laser cutting of carbon steel. J Mater Sci 29:5607–5611
39. Powell J, Petring D, Kumar RV, Al-Mashikhi SO, Kaplan AFH, Voisey KT (2009) Laser-oxygen cutting of mild steel: the thermodynamics of the oxidation reaction. J Phys D Appl Phys 42:015504
40. Geiger M, Bergmann HW, Nuss R (1988) Laser cutting of steel sheets. Proc SPIE 1022:20–33
41. Chen S-L (1998) The effects of gas composition on the CO_2 laser cutting of mild steel. J Mater Process Technol 73:147–159
42. Powell J, Ivarson A, Magnusson C (1993) Laser cutting of steels: a physical and chemical analysis of the particles ejected during cutting. J Laser Appl 5:25–31
43. Poprawe R, König W (2001) Modeling, monitoring and control in high quality laser cutting. CIRP Ann—Manuf Technol 50:137–140
44. Yudin P, Kovalev O (2009) Visualization of events inside kerfs during laser cutting of fusible metal. J Laser Appl 21:39–45
45. Hirano K, Fabbro R (2011) Experimental investigation of hydrodynamics of melt layer during laser cutting of steel. J Phys D Appl Phys 44:105502
46. Riveiro A, Quintero F, Lusquiños F, Comesaña R, Pou J (2011) Study of melt flow dynamics and influence on quality for CO_2 laser fusion cutting. J Phys D Appl Phys 44:135501
47. Chen S-L (1997) In-process monitoring of the cutting front of CO_2 laser cutting with off-axis optical fibre. Int J Adv Manuf Technol 13:685–691
48. Schuöcker D, Abel W (1984) Material removal mechanism of laser cutting. Proc SPIE 455:88–95
49. Schuöcker D (1986) Dynamic phenomena in laser cutting and cut quality. Appl Phys B 40:9–14
50. Vicanek M, Simon G (1987) Momentum and heat transfer of an inert gas jet to the melt in laser cutting. J Phys D Appl Phys 20:1191–1196
51. Schulz W, Simon G, Urbassek HM, Decker I (1987) On laser fusion cutting of metals. J Phys D Appl Phys 20:481–488
52. Schulz W, Kostrykin V, Nießen M, Michel J, Petring D, Kreutz EW, Poprawe R (1999) Dynamics of ripple formation and melt flow in laser beam cutting. J Phys D Appl Phys 32:1219–1228
53. Vicanek M, Simon G, Urbassek HM, Decker I (1987) Hydrodynamical instability of melt flow in laser cutting. J Phys D Appl Phys 20:140–145

54. Wandera C, Kujanpaa V (2010) Characterization of the melt removal rate in laser cutting of thick-section stainless steel. J Laser Appl 22:62–70
55. Hirano K, Fabbro R (2011) Experimental observation of hydrodynamics of melt layer and striation generation during laser cutting of steel. Phys Procedia 12(Part A):555–564
56. Powell J, Al-Mashikhi SO, Kaplan AFH, Voisey KT (2011) Fibre laser cutting of thin section mild steel: an explanation of the 'striation free' effect. Opt Lasers Eng 49:1069–1075
57. Schober A, Musiol J, Daub R, Feil J, Zaeh MF (2012) Experimental investigation of the cutting front angle during remote fusion cutting. Phys Procedia 39:204–212
58. Di Pietro P, Yao YL (1995) A numerical investigation into cutting front mobility in CO_2 laser cutting. Int J Mach Tool Manuf 35:673–688
59. Ermolaev GV, Kovalev OB, Orishich AM, Fomin VM (2006) Mathematical modelling of striation formation in oxygen laser cutting of mild steel. J Phys D Appl Phys 39:4236–4244
60. Li L, Sobih M, Crouse PL (2007) Striation-free laser cutting of mild steel sheets. CIRP Ann—Manuf Technol 56:193–196
61. Schuöcker D (1988) Heat conduction and mass transfer in laser cutting. Proc SPIE 952:592–599
62. Tsai MJ, Weng CI (1993) Linear stability analysis of molten flow in laser cutting. J Phys D Appl Phys 26:719–727
63. Schulz W, Simon G, Vicanek M, Decker I (1987) Influence of the oxidation process in laser gas cutting. Proc SPIE 801:331–336
64. Powell J, Ivarson A, Ohlsson L, Magnusson C (2000) Conductive losses experienced during CO_2 laser cutting. High Temper Mater Process 4:201–211
65. Onuseit V, Ahmed MA, Weber R, Graf T (2011) Space-resolved spectrometric measurements of the cutting front. Phys Procedia 12(Part A):584–590
66. Olsen FO (1994) Fundamental mechanisms of cutting front formation in laser cutting. Proc SPIE 2207:402–413
67. Chen K, Lawrence Yao Y (1999) Striation formation and melt removal in the laser cutting process. J Manuf Process 1:43–53
68. Heidenreich B, Jüptner W, Sepold G (1996) Fundamental investigations of the burn-out phenomenon of laser cut edges. Lasers Eng 5:1–10
69. Wee LM, Li L (2005) An analytical model for striation formation in laser cutting. Appl Surf Sci 247:277–284
70. Duan J, Man HC, Yue TM (2001) Modelling the laser fusion cutting process: III. Effects of various process parameters on cut kerf quality. J Phys D Appl Phys 34:2143–2150
71. Chen K, Lawrence Yao Y, Modi V (2001) Gas dynamic effects on laser cut quality. J Manuf Process 3:38–49
72. Ermolaev GV, Kovalev OB (2009) Simulation of surface profile formation in oxygen laser cutting of mild steel due to combustion cycles. J Phys D Appl Phys 42:185506
73. Kovalev OB, Yudin PV, Zaitsev AV (2008) Formation of a vortex flow at the laser cutting of sheet metal with low pressure of assisting gas. J Phys D Appl Phys 41:155112
74. Farooq K, Kar A (1998) Removal of laser-melted material with an assist gas. J Appl Phys 83:7467–7473
75. Kovalev OB, Yudin PV, Zaitsev AV (2009) Modeling of flow separation of assist gas as applied to laser cutting of thick sheet metal. Appl Math Model 33:3730–3745
76. Karatas C, Keles O, Uslan I, Usta Y (2006) Laser cutting of steel sheets: influence of workpiece thickness and beam waist position on kerf size and stria formation. J Mater Process Technol 172:22–29
77. Kaebernick H, Bicleanu D, Brandt M (1999) Theoretical and experimental investigation of pulsed laser cutting. CIRP Ann—Manuf Technol 48:163–166
78. Decker I, Ruge J, Atzert U (1984) Physical models and technological aspects of laser gas cutting. Proc SPIE 455:81–87
79. Sobih M, Crouse PL, Li L (2008) Striation-free fibre laser cutting of mild steel sheets. Appl Phys A 90:171–174

80. Ledenev VI, Karasev VA, Yakunin VP (1999) On cyclical mechanism of kerf formation under gas assisted laser cutting of metals. Proc SPIE 3688:157–162
81. Di Pietro P, Yao YL (1995) A new technique to characterize and predict laser cut striations. Int J Mach Tool Manuf 35:993–1002
82. Schuöcker D, Aichinger J, Majer R (2012) Dynamic phenomena in laser cutting and process performance. Phys Procedia 39:179–185
83. Makashev NK, Asmolov ES, Blinkov VV, Boris AY, Burmistrov AV, Buzykin V, Makarov VA (1994) Gas-hydro-dynamics of CW laser cutting of metals in inert gas. Proc SPIE 2257:2–9
84. King TG, Powell J (1986) Laser-cut mild steel—factors affecting edge quality. Wear 109:135–144
85. Kaebernick H, Jeromin A, Mathew P (1998) Adaptive control for laser cutting using striation frequency analysis. CIRP Ann—Manuf Technol 47:137–140
86. Yan Y, Li L, Sezer K, Whitehead D, Ji L, Bao Y, Jiang Y (2012) Nano-second pulsed DPSS Nd: YAG laser striation-free cutting of alumina sheets. Int J Mach Tool Manuf 53:15–26
87. Dahotre NB, Harimkar SP (2008) Laser Fabrication and Machining of Materials. Springer, New York
88. Arata Y, Maruo H, Miyamoto I, Takeuchi S (1981) Quality in laser-gas-cutting stainless steel and its improvement. Trans Jpn Weld Res Inst 10:129–139
89. Nielsen SE (1997) Developments in laser beam cutting of thick materials. Industrial Laser Review 12: 11–13
90. Chen S-L (1999) The effects of high-pressure assistant-gas flow on high-power CO_2 laser cutting. J Mater Process Technol 88:57–66
91. O'Neill W, Steen WW (1995) A three-dimensional analysis of gas entrainment operating during the laser-cutting process. J Phys D Appl Phys 28:12–18
92. Sheng PS, Joshi VS (1995) Analysis of heat-affected zone formation for laser cutting of stainless steel. J Mater Process Technol 53:879–892
93. Shanjin L, Yang W (2006) An investigation of pulsed laser cutting of titanium alloy sheet. Opt Lasers Eng 44:1067–1077
94. Riveiro A, Quintero F, Lusquiños F, Comesaña R, Pou J (2010) Influence of assist gas nature on the surfaces obtained by laser cutting of Al–Cu alloys. Surf Coat Technol 205:1878–1885
95. Riveiro A, Quintero F, Lusquiños F, Comesaña R, Pou J (2010) Parametric investigation of CO2 laser cutting of 2024–T3 alloy. J Mater Process Technol 210:1138–1152
96. Quintero F, Pou J, Lusquiños F, Boutinguiza M, Soto R, Pérez-Amor M (2004) Quantitative evaluation of the quality of the cuts performed on mullite-alumina by Nd:YAG laser. Opt Lasers Eng 42:327–340
97. Rao BT, Kaul R, Tiwari P, Nath AK (2005) Inert gas cutting of titanium sheet with pulsed mode CO_2 laser. Opt Lasers Eng 43:1330–1348
98. Scintilla LD, Tricarico L, Wetzig A, Mahrle A, Beyer E (2011) Primary losses in disk and CO_2 laser beam inert gas fusion cutting. J Mater Process Technol 211:2050–2061
99. Powell J, Ivarson A, Kamalu J, Broden G, Magnusson C (1993) Role of oxygen purity in laser cutting of mild steel. Proc SPIE 1990:433–442
100. Belic I, Stanic J (1987) A method to determine the parameters of laser iron and steel cutting. Opt Laser Technol 19:309–311
101. Belić I (1989) A method to determine the parameters of laser cutting. Opt Laser Technol 21:277–278
102. Chryssolouris G (1991) Laser Machining, Therory and Practice. Springer, New York
103. Lepore M, Dell'Erba M, Esposito C, Daurelio G, Cingolani A (1983) An investigation of the laser cutting process with the aid of a plane polarized CO_2 laser beam. Opt Lasers Eng 4:241–251
104. Muys P, Youn M (2008) Mathematical Modeling of Laser Sublimation Cutting. Laser Phys 18:495–499

105. Niziev V, Nesterov A (1999) Influence of beam polarization on laser cutting efficiency. J Phys D Appl Phys 32:1455–1461
106. Scintilla LD, Tricarico L, Mahrle A, Wetzig A, Beyer E (2012) A comparative study of cut front profiles and absorptivity behavior for disk and CO_2 laser beam inert gas fusion cutting. J Laser Appl 24:052006
107. Poprawe R, Schulz W, Schmitt R (2010) Hydrodynamics of material removal by melt expulsion: perspectives of laser cutting and drilling. Phys Procedia 5(Part A):1–18
108. Olsen FO, Hansen KS, Nielsen JS (2009) Multibeam fiber laser cutting. J Laser Appl 21:133–138
109. Petring D, Molitor T, Schneider F, Wolf N (2012) Diagnostics, modeling and simulation: three keys towards mastering the cutting process with fiber, disk and diode lasers. Phys Procedia 39:186–196
110. Wandera C, Kujanpää V, Salminen A (2011) Laser power requirement for cutting thick-section steel and effects of processing parameters on mild steel cut quality. Proc Inst Mech Eng B J Eng Manuf 225:651–661
111. Sparkes M, Gross M, Celotto S, Zhang T, O'Neill W (2008) Practical and theoretical investigations into inert gas cutting of 304 stainless steel using a high brightness fiber laser. J Laser Appl 20:59–67
112. Grevey DF, Desplats H (1994) Comparison of the performance obtained with a YAG laser cutting according to the source operation mode. J Mater Process Technol 42:341–348
113. Ivarson A, Powell J, Magnusson C (1996) The role of oxygen pressure in laser cutting mild steels. J Laser Appl 8:191–196
114. Thawari G, Sundar JKS, Sundararajan G, Joshi SV (2005) Influence of process parameters during pulsed Nd:YAG laser cutting of nickel-base superalloys. J Mater Process Technol 170:229–239
115. Bicleanu D, Brandt M, Kaebernick H (1996) An analytical model for pulsed laser cutting of metals. In: Proceedings 15th international congress on applications of lasers and electro-optics, 68–77
116. Lee CS, Gael A, Osada H (1985) Parametric studies of pulsed-laser cutting of thin metal plates. J Appl Phys 58:1339–1343
117. Pfeifer R, Herzog D, Hustedt M, Barcikowski S (2010) Pulsed Nd:YAG laser cutting of NiTi shape memory alloys—Influence of process parameters. J Mater Process Technol 210:1918–1925
118. Powell J, Tan WK, Maclennan P, Rudd D, Wykes C, Engstrom H (2000) Laser cutting stainless steel with dual focus lenses. J Laser Appl 12:224–231
119. Rocca AVL, Borsati L, Cantello M (1994) Nozzle design to control fluid-dynamics effects in laser cutting. Proc SPIE 2207:354–368
120. Powell J, Frass K, Menzies IA (1987) 2.5 kW Laser cutting of steels; Factors affecting cut quality in sections up to 20 mm. Proc SPIE 801:278–282
121. Man HC, Duan J, Yue TM (1997) Design and characteristic analysis of supersonic nozzles for high gas pressure laser cutting. J Mater Process Technol 63:217–222
122. Leidinger D, Penz A, Schuöcker D (1995) Improved manufacturing processes with high power lasers. Infrared Phys Technol 36:251–266
123. Quintero F, Pou J, Lusquiños F, Boutinguiza M, Soto R, Pérez-Amor M (2003) Comparative study of the influence of the gas injection system on the Nd:yttrium-aluminum-garnet laser cutting of advanced oxide ceramics. Rev Sci Instrum 74:4199–4205
124. Na SJ, Yang YS, Koo HM, Kim TK (1989) Effect of shielding gas pressure in laser cutting of sheet metals. Trans ASME J Eng Mater Technol 111:314–318
125. Quintero F, Pou J, Fernández JL, Doval AF, Lusquiños F, Boutinguiza M, Soto R, Pérez-Amor M (2006) Optimization of an off-axis nozzle for assist gas injection in laser fusion cutting. Opt Lasers Eng 44:1158–1171
126. Brandt AD, Settles GS (1997) Effect of nozzle orientation on gas dynamics of inert gas laser cutting of mild steel. J Laser Appl 9:269–277
127 Chryssolouris G, Choi WC (1989) Gas jet effects on laser cutting. Proc SPIE 1042:86-96

128. Hsu MJ, Molian PA (1995) Off-axial, gas-jet-assisted, laser cutting of 6.35-mm thick stainless steel. Trans ASME J Eng Ind 117:272–276
129. Prasad GVS, Siores E, Wong WCK (1998) Laser cutting of metallic coated sheet steels. J Mater Process Technol 74:234–242
130. Arai T, Riches S (1997) Thick plate cutting with spining laser beam. In: Proceedings 16th international congress on applications of lasers and electro-optics, 19–26
131. O'Neill W, Gabzdy JT (2000) New developments in laser-assisted oxygen cutting. Opt Lasers Eng 34:355–367
132. Ghany KA, Newishy M (2005) Cutting of 1.2 mm thick austenitic stainless steel sheet using pulsed and CW Nd:YAG laser. J Mater Process Technol 168:438–447
133. Samant AN, Dahotre NB (2009) Laser machining of structural ceramics—A review. J Europ Ceram Soc 29:969–993
134. Yan Y, Li L, Sezer K, Whitehead D, Ji L, Bao Y, Jiang Y (2011) Experimental and theoretical investigation of fibre laser crack-free cutting of thick-section alumina. Int J Mach Tool Manuf 51:859–870
135. Lu G, Siores E, Wang B (1999) An empirical equation for crack formation in the laser cutting of ceramic plates. J Mater Process Technol 88:154–158
136. Black I, Chua KL (1997) Laser cutting of thick ceramic tile. Opt Laser Technol 29:193–205
137. Black I, Livingstone SAJ, Chua KL (1998) A laser beam machining (LBM) database for the cutting of ceramic tile. J Mater Process Technol 84:47–55
138. Lei H, Lijun L (1999) A study of laser cutting engineering ceramics. Opt Laser Technol 31:531–538
139. Wang J, Wong WCK (1999) CO_2 laser cutting of metallic coated sheet steels. J Mater Process Technol 95:164–168
140. De Graaf RF, Meijer J (2000) Laser cutting of metal laminates: analysis and experimental validation. J Mater Process Technol 103:23–28
141. Yilbas BS, Khan S, Raza K, Keles O, Ubeyli M, Demir T, Karakas MS (2010) Laser cutting of 7,050 Al alloy reinforced with Al_2O_3 and B_4C composites. Int J Adv Manuf Technol 50:185–193
142. Kagawa Y, Utsunomiya S, Kogo Y (1989) Laser cutting of CVD-SiC fibre/A6061 composite. J Mater Sci Lett 8:681–683
143. Cenna AA, Mathew P (1997) Evaluation of cut quality of fibre-reinforced plastics—A review. Int J Mach Tool Manuf 37:723–736
144. Mello MD (1986) Laser cutting of non-metallic composites. Proc SPIE 668:288–290
145. Caprino G, Tagliaferri V (1988) Maximum cutting speed in laser cutting of fiber reinforced plastics. Int J Mach Tool Manuf 28:389–398
146. Ready JF (1997) Industrial Applications of Lasers. Academic Press, San Diego
147. Migliore L (1996) Laser-material interactions. In: Migliore L (ed) Laser Materials Processing. Marcel Dekker, Inc., New York
148. Rofin (2004) Introduction to industrial laser materials processing. http://www.obrusn.torun.pl/htm0/prod_images/rofin/Laserbook.pdf. Accessed 23 August 2012
149. Olsen FO (2011) Laser cutting from CO_2 laser to disk or fiber laser—possibilities and challenges. In: Proceedings 30th international congress on applications of lasers and electro-optics, Paper #101
150. Fieret J, Terry MJ, Ward BA (1986) Aerodynamic interactions during laser cutting. Proc SPIE 668:53–62
151. Fieret J, Terry MJ, Ward BA (1987) Overview of flow dynamics in gas-assisted laser cutting. Proc SPIE 801:243–250
152. Molian PA (1993) Dual-beam CO_2 laser cutting of thick metallic materials. J Mater Sci 28:1738–1748
153. Toyserkani E, Khajepour A, Corbin S (2005) Laser Cladding. CRC Press LLC, Boca Raton
154. Nath AK (2013) High power lasers in material processing applications: an overview of recent development. In: Majumdar JD, Manna I (eds) Laser-assisted fabrication of materials. Springer-Verlag, Berlin

155. Steen WM, Mazumder J (2010) Laser Material Processing. Springer, London
156. Powell J, Frass K, Menzies IA, Fuhr H (1988) CO_2 laser cutting of non-ferrous metals. Proc SPIE 1020:156–163
157. Riveiro A, Quintero F, Lusquiños F, Comesaña R, del Val J, Pou J (2011) The role of the assist gas nature in laser cutting of aluminum alloys. Phys Procedia 12(Part A): 548–554

Chapter 3
CO_2 Laser Cutting of Triangular Geometry in Aluminum Foam

B. S. Yilbas, S. S. Akhtar and O. Keles

Abstract In laser cutting process, thermal stress is formed around the cut edges, which becomes important for the geometries having the corners. In the present study, laser cutting of triangular shape in aluminum foam is carried out. Temperature and stress fields are predicted using a finite element method in line with experimental conditions. Metallurgical and morphological changes in the cut sections are examined incorporating scanning electron microscope (SEM) and energy-dispersive spectroscopy. It is found that the cut sections are free from defects such as sideways burning and large-scale burrs. von Mises stress is high in the corners of the triangular-shaped laser cut workpieces.

3.1 Introduction

Laser cutting of porous materials have several advantages over the conventional cutting methods. Some of these advantages are no mechanical contact between the workpiece and the cutting tool, precision of operation, high speed processing, and low cost. In conventional machining, the mechanical forces between the cutting tool and workpiece can cause local damages in the cut section of the foam material, which in turn reduces the quality of the end product. In the absence of mechanical force during laser cutting, the local damages due to mechanical forces can be eliminated. However, excessive heating during laser cutting results in defect sites, such as sideways burning and dross attachments in the cut sections.

B. S. Yilbas (✉) · S. S. Akhtar
Mechanical Engineering Department,
King Fahd University of Petroleum & Minerals, Dhahran, Saudi Arabia
e-mail: bsyilbas@kfupm.edu.sa

O. Keles
Mechanical Engineering Department, Gazi University, Ankara, Turkey

J. P. Davim (ed.), *Nontraditional Machining Processes*,
DOI: 10.1007/978-1-4471-5179-1_3, © Springer-Verlag London 2013

This is particularly true for laser cutting of geometries having the sharp corners. In addition, high temperature gradients cause the formation of high thermal stresses at edges of the cut sections while limiting the practical applications of the machined parts. Consequently, investigation into laser cutting of a metallic foam and determination of thermal stress development becomes essential.

Considerable research study has been carried out to examine laser processing of metallic foams. Laser-assisted aluminum foaming was studied by Kathuria [1]. He demonstrated that the laser interaction time had significant effect on the foam cell morphology and the density. Laser processing of metal foam was investigated by Schmid et al. [2]. They demonstrated that the laser processing enabled to restructure the foam material similar to that observed in the bone structure. Nd:YAG laser machining of foam was examined by Fujimura et al. [3]. They reported the relationships between the laser power and the cutting speed for optimum processing with minimum defects. Laser processing of aluminum foam–filled tubes was investigated by Campana et al. [4]. They demonstrated that the laser could be used as an effective tool to machine the foam parts. Laser produced metal foam was studied by Carcel et al. [5]. They showed that the laser produced aluminum and titanium foams had closed-cell structures. Laser-induced coating of foam on aluminum alloy was examined by Ocelik et al. [6]. Their findings revealed that the tensile behavior of the foam layer was close to the behavior of bulk aluminum than the conventional closed-cell foams. Laser drilling of holes into foam was investigated by Yoshida et al. [7]. They presented the effect of laser pulse width on the hole structure and the pileup around the hole in a single-pulse drilling. Laser processing aluminum panels was studied by Guglielmotti et al. [8]. They indicated that laser scanning speed and laser power were the most important parameters affecting the end product quality.

Although laser processing of metal foam was investigated previously [9], the focus of the study was assessing the end product quality identifying the optimum process parameters. Therefore, the thermal analysis associated with thermal stress in the machining sections was left obscure. In the present study, laser cutting of triangular shape in aluminum foam is considered and thermal stress analysis, including temperature and stress fields, is carried out around the cut edges using the ABAQUS finite element code [10]. Experiment is conducted to examine the morphological and metallurgical changes around the cut edges.

3.2 Experimental

A CO_2 laser (LC-ALPHAIII) delivering nominal output power of 2 kW was used to irradiate the workpiece surface. Focused laser beam has a Gaussian intensity distribution at the irradiated surface with a Gaussian parameter 'a' of 0.0002 m. Nitrogen assisting gas was used during the cutting process. To identify the optimum cutting parameters, several cutting tests were carried out. The criteria introduced for the optimization tests were based on the less sideways burning

Table 3.1 Laser cutting conditions used in the experiment

Scanning speed (cm/s)	Power (W)	Frequency (Hz)	Nozzle gap (mm)	Nozzle diameter (mm)	Focus setting (mm)	N$_2$ pressure (kPa)
10	1,800	1,500	1.5	1.5	127	500

Table 3.2 Elemental composition at three locations on the kerf surface (wt %)

Spectrum	O	Si	Al
Spectrum 1	25	9	Balance
Spectrum 2	29	10	Balance
Spectrum 3	21	10	Balance

around the kerf edges, small dross attachment at the kerf exit, and low striations at the kerf surface. It was observed that increasing laser power to 10 % from its optimum value, the kerf width size increased almost 25 % and dross attachment at the kerf exit was also observed. However, increasing laser scanning speed 10 %, irregular cuts were resulted such that the kerf width became a wedge-like shape rather than the parallel sided. Therefore, the optimum laser cutting parameters were obtained prior to simulations, and they are given in Table 3.1.

Aluminum foam plates used in the cutting experiments were 5 mm thick, and they had closed-cell structure with the porosity of the order of 78 %. JEOL JDX-3530 scanning electron microscope (SEM) was used to obtain photomicrographs of the cut sections. The geometry cut into the aluminum foam is equilateral triangle with one edge is 7 mm. EDS analysis was carried out for different locations at the kerf surface, and the EDS data are given in Table 3.2. The error related to the EDS analysis was on the order of 3 %.

3.3 Thermal Stress Analysis

Figure 3.1 shows the schematic view of the laser cutting and the coordinate system. Since the foam material contains large size air cells, the governing equation of energy transfer for laser cutting is modified to incorporate the presence of pores in the workpiece. The continuum conservation equations are used in the simulations with the assumption of effective properties for the aluminum foam in line with the previous studies [11, 12]. In the analysis, the gas dynamics effect of the assisting gas on the cutting process is omitted due to the complicated nature of the gas flow behavior in the large cavities in the foam; however, it is referred to the previous studies [13–16]. The convective cooling effect of the assisting gas in the cutting section is incorporated in the analysis through the convection boundary conditions.

The thermal analysis is conducted incorporating the temperature dependence of the material properties, namely the density, specific heat, and thermal conductivity. The values of the effective properties are calculated for the lotus-type

Fig. 3.1 Schematic view of laser cut geometry in aluminum foam and the directions of the laser cut

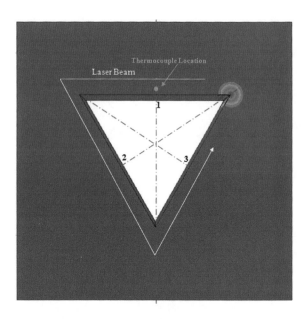

porous metal through incorporating the rule of mixture [12]. Therefore, the effective properties are [12]

$$\rho_{\text{eff}}(T) = (1 - \varepsilon)\rho_s(T) + \varepsilon\rho_f(T) \tag{3.1}$$

and

$$C_{\text{peff}}(T) = \frac{(1 - \varepsilon)\rho_s(T)C_{ps}(T) + \varepsilon\rho_f(T)C_{pf}(T)}{(1 - \varepsilon)\rho_s(T) + \varepsilon\rho_f(T)} \tag{3.2}$$

and

$$k_{\text{eff}}(T) = \frac{(1 - \varepsilon)\rho_s(T)k_s(T) + \varepsilon\rho_f(T)k_f(T)}{(1 - \varepsilon)\rho_s(T) + \varepsilon\rho_f(T)} \tag{3.3}$$

where $\rho_{\text{sub}}(T)$, $C_{\text{psub}}(T)$, and $k_{\text{eff}}(T)$ are the density and specific heat of porous metal, respectively, and ε is the porosity expressed by the volume ratio of pores versus the total volume of the porous metal. The subscripts (sub) 'eff,' 's,' and 'f' indicate the effective property, property of the nonporous aluminum metal, and property of the gas in the pores, respectively.

To analyze the phase change problem, the enthalpy method is used [10]. The specific heat capacity is associated with the internal energy gain of the substrate material, that is, $Cp(T) = \frac{\partial U}{\partial T}$. Latent heat effects are significant in the laser melting of aluminum foam material due to phase change and must be taken into account. The equivalent latent heat ($L_{\text{eq}} = \frac{L_s}{1 + \frac{\rho_f}{\rho_s}\varepsilon\left(1 + \frac{\varepsilon}{\varepsilon - 1}\right)}$, where L_{eq} is the equivalent latent heat of melting, and L_s is the melting heat of aluminum) is used in terms of solidus and liquidus temperatures (the lower and upper temperature bounds of the phase change range) and the total internal energy associated with the phase change.

The metal foam behavior is different from that of the solid metal, and the study conducted by Deshpande and Fleck [17] is considered. Since they developed a three-dimensional model based on the experimental tests of aluminum foam, which has been built in the finite element package ABAQUS [10], the model assumes similar behaviors in tension and compression, hence incorporating the isotropic hardening. Therefore, in the present work, three-dimensional finite element analysis was carried out using the dynamic explicit code ABAQUS/Explicit incorporating the crushable foam model with the isotropic hardening. The stress analysis was coupled with the previous thermal analysis, which was carried out for the laser treatment process. Moreover, the crushable foam plasticity model was based on the assumption that the resulting deformation was not recoverable instantaneously and could be idealized as plastic for short duration events. The contribution of the mean stress on the yield function was realized through a material parameter known as a shape factor. It defined the aspect ratio of the elliptical stress. The shape factor quantitatively distinguished the plastic behavior of metal foams from solid metals [17]. The detail of the analysis is referred to the previous study [17].

During the laser cutting process, self-annealing took place in the previously cut sections due to heat transfer. In order to simulate this situation, the consideration was made such that the relaxation of stresses and plastic strains occurred as the workpiece was heated to above melting temperature during the laser processing and then it cooled to a room temperature. In this case, temperature-dependent foam hardening sub-option was used in the simulations and low values of yield stress were assumed above the melting temperature. The temperature-dependent properties, which are used in the simulations, are given in Table 3.3.

Table 3.3 Properties of aluminum used in the simulations [18]

Temperature (K)	Effective specific heat (J/kg K)	Effective thermal conductivity (W/m K)	Yield stress (MPa)
300	870	9.07	1.0
400	920	9.16	0.8
500	970	9.09	0.65
600	1,000	9.15	0.49
700	1,030	9.69	0.26
800	1,050	9.93	0.24
823	1,050	9.93	0.12
893	1,050	9.93	0.015

Effective density (ρ_{eff}) = 250 kg/m^3 , latent heat = 370,000 J/kg, solidus temperature = 835 K, liquidus temperature = 865 K. Expansion coefficient = 2.34 $\times$ 10^{-5} 1/K, Poisson's ratio = 0.25
Crushable foam properties: Compression yield stress ratio = 1, plastic Poisson's ratio = 0.25

3.4 Results and Discussion

Laser cutting of triangular geometry into aluminum foam is carried out. Temperature and thermal stress fields are predicted in line with experimental conditions.

Figure 3.2 shows temperature distribution at the top and bottom perimeters of the triangular cut edges for different cooling periods, while Fig. 3.3 shows temperature contours in the cut sections. The cooling period starts immediately after the completion of cutting, which corresponds to t = 0.21 s. Temperature is high across the irradiated spot at the onset of cooling cycle (t = 0.21 s). In addition, temperature exceeds the liquidus temperature of the substrate material indicating the superheating of liquid phase at the central region of the irradiated spot. It is observed that the temperature remains within the range of solidus and liquidus temperatures in the region close to the irradiated spot edge. This represents the mushy zone in this region despite the fact that the size of this region is small.

Fig. 3.2 Temperature profiles along the circumference of the cut sections at the *top* and *bottom* edges for different cooling periods

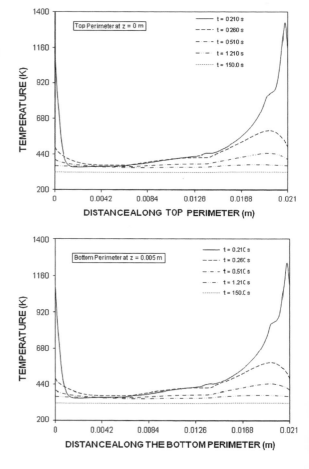

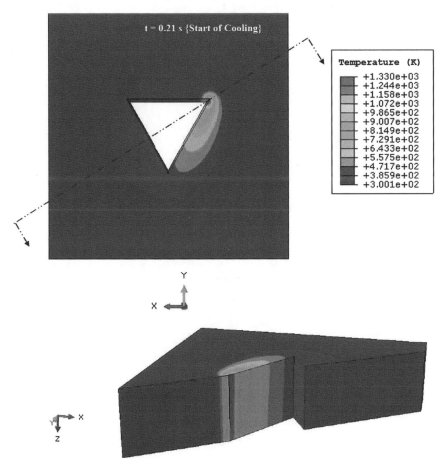

Fig. 3.3 Temperature contours in the cut section onset of cooling period initiation

Temperature decays gradually behind the irradiated spot around perimeter of the cut edges. This is attributed to the slow cooling of the initially cut edges due to convection and conduction heat transfer. As the cooling period progresses, temperature decays sharply around the perimeter and the cooling cycle ends after t = 100 s; in which case, temperature around the cut edges reduces to initial temperature. When comparing temperature behavior at the top and bottom perimeters of the triangular cut edges, temperature profiles show similar behavior in both cases. However, the values of temperature at the bottom edge are lower than that corresponding to the top edge. This is associated with the absorption of the laser beam prior to reaching to the bottom edge, which results in less temperature raise than that at the top surface.

Figure 3.4 shows von Mises stress distribution along the top and bottom perimeters of the triangular cut in the aluminum foam for different cooling periods, while Fig. 3.5 shows von Mises stress contours in the cut sections. von Mises stress

Fig. 3.4 von Mises stress
profiles along the
circumference of the cut
sections at the *top* and *bottom*
edges for different cooling
periods

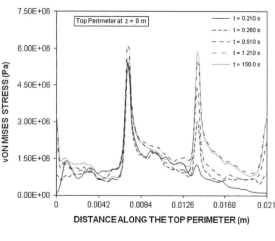

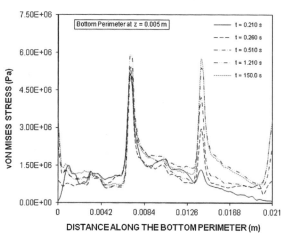

attains high values in the regions close to the corners of the cut edges. This
indicates that the thermal expansion in this region is suppressed by the presence of
corners giving rise to the attainment of high stress levels in these regions. In
addition, von Mises stress reduces sharply along the perimeters of the cut edges as
the distance increases from the corner regions. This is attributed to the free
expansion of the top and bottom edges, which do not suffer from the boundary
constraints. The rise of von Mises stress reaching to its maximum is sharper than
that of its fall from the maximum value. This behavior is attributed to the temperature gradient, which results in the sharp rise of von Mises stress in the corner
regions. When comparing von Mises stress behavior along the top and bottom
perimeters, both behave in a similar manner; however, the maximum von Mises
stress occurs at the top edge. This is attributed to the attainment of relatively
higher temperature gradients at the top perimeter when compared to that corresponding to the bottom edge. The maximum value of von Mises stress is on the

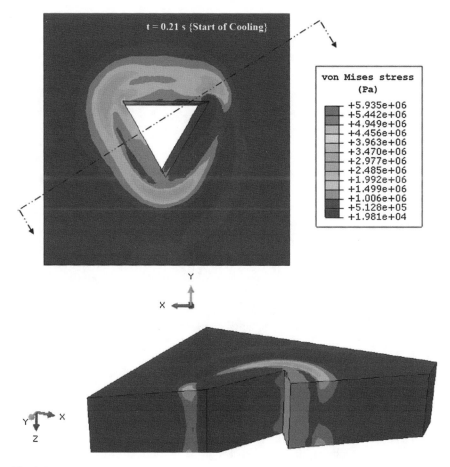

Fig. 3.5 von Mises stress contours in the cut section onset of cooling period initiation

order of 6 MPa after the end of cooling cycle, which is less than the yielding limit of the substrate material. Therefore, the crack formation along the perimeters of the cut edges is less likely after the completion of laser cutting.

Figure 3.6 shows temperature distribution along the z-axis at three locations around the perimeter of the cut edges at the onset of cooling cycle (t = 0.21 s). Temperature decays almost gradually along the z-axis at all three locations as shown in Fig. 3.1 provided that the decay is more pronounced at the location # 3 where the laser beam is ceases at the end of cutting. It should be noted that the locations around the perimeter at z = 0 represents the free surface of the workpiece where the laser beam is incident onto it and at z = 0.005 m corresponds to the bottom surface of the workpiece where the laser beam exits the kerf. The gradual decay of temperature along the z-axis is associated with the absorption of the laser beam along the thickness of the workpiece during the cutting process. Temperature reduces notably at locations away from the last location of the laser

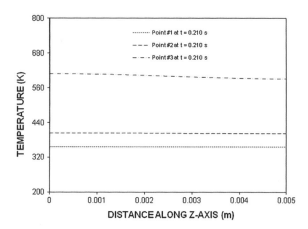

Fig. 3.6 Temperature profiles along the z-axis at different locations along the circumference of the cut section for t = 0.21 s of the cooling period

beam (location # 3) around the perimeter. This indicates the occurrence of the high cooling rates along the edges of the cut section; in which case, convection and conduction heat transfer at the kerf surface is responsible for the attainment of the high cooling rates.

Figure 3.7 shows von Mises stress distributions along the z-axis at three locations along the perimeters of the triangular cut shape at the end of cooling cycle (t = 150 s). von Mises stress attains low values in the region close to the top free surface (z = 0 m) and at the bottom surface (z = 0.005 m) at the end of cooling cycle. The attainment of low stresses in the top and bottom cut surfaces is associated with the free expansion of the surfaces in these regions. It should be noted that the top and bottom surfaces of the workpiece are free to expand. Due to the constraint on thermal expansion at the mid-thickness of the cut section, von Mises stress attains high values in this region. In this case, the suppression of the free expansion results in high stress levels at the mid-thickness of the workpiece. Moreover, von Mises stress attains low values at location where the laser beam is ceased off (location # 3). This is attributed to the self-annealing effect of the

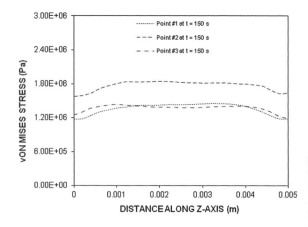

Fig. 3.7 von Mises stress profiles along the z-axis at different locations along the circumference of the cut section for t = 150 s of the cooling period

initially cut edges. Consequently, heat transfer from the initially cut edges toward the location # 3 results in low cooling rates from this location while lowering the stress levels at this location at the end of cooling cycle.

Figure 3.8 shows optical photographs of the top and bottom surfaces of the triangular cut shape. It can be observed that no excessive heating takes place at the cut edges; consequently, no large-scale sideways burning is observed around the perimeter of the cut section. In addition, no clear striation pattern is formed along the cut edges. This is attributed to the proper selection of the cutting parameters, which results in high-quality cutting. Fig. 3.9 shows SEM micrographs of the laser cut sections. It can be observed that the cut surfaces are free from cracks and the voids in the foam structure are not covered by the molten

Fig. 3.8 Optical photographs of *top* and *bottom* of the laser cut geometry

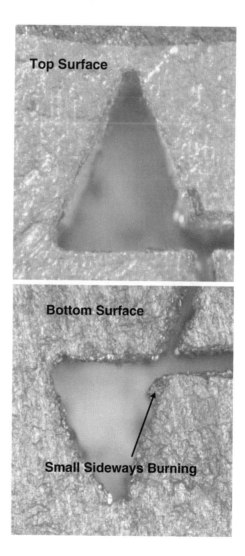

Fig. 3.9 SEM micrographs
of laser kerf surface

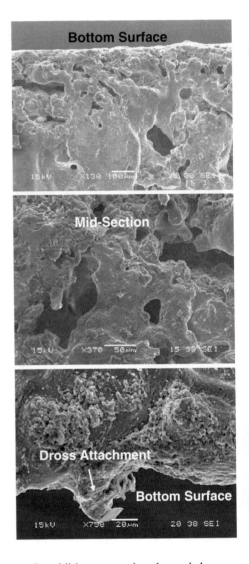

Fig. 3.9 SEM micrographs of laser kerf surface

flow generated during the cutting process. In addition, excessive thermal damage due to high temperature oxidation reaction is not observed around the cut edges. This is associated with the use of nitrogen assisting gas, which prevents the oxidation reactions taking place in the cut section. However, small amount of dross attachments is observed at the kerf exit. The dross attachment is local and does not form continuous deposits around the perimeter of the cut section. The occurrence of local dross attachment is attributed to the molten flow from the cut section. In this case, assisting gas loses its axial momentum in the kerf due to the fluid friction generated in the narrow kerf width and the drag force generated at the molten surface reduces toward the kerf exit. The rate of molten flow ejection due to drag force reduces at the kerf exit. Hence, the surface tension effect of the retarded molten flow at the kerf exit contributes to the formation of the dross attachment at the kerf exit. Table 3.2 gives the EDS data at the kerf surface. It is evident that the elemental composition of the workpiece remains the same at different locations at the kerf surface. However, the presence of oxygen suggests that low rate oxidation reactions took place during the cutting process. This is associated with the presence of air in the voids of the aluminum foam prior to cutting, which releases during the cutting process while resulting in oxide formations at the resolidified molten surface. It should be noted that the oxidation reactions are exothermic and give rise to generation of excessive heat at the kerf surface. Since excessive thermal damage is not observed at the kerf surface, the oxidation reactions take place at a low scale and they do not result in excessive heat generation in the cutting sections.

3.5 Conclusion

Laser cutting of triangular geometry into aluminum foam is investigated. Thermal stress field developed around the cut section is predicted using the finite element code (ABAQUS) in line with the experimental conditions. The morphological and metallurgical changes at the kerf surface are examined incorporating electron scanning microscope and energy-dispersive spectroscopy. Temperature well exceeds the liquidus temperature of the substrate material at the irradiated center and small size of the mushy zone is observed in the vicinity of the irradiated spot edge. von Mises stress attains high values in the region of the corners of the triangle cut geometry. This is attributed to the constraint against the thermal expansion of the cut section in the corner region. von Mises stress becomes low at the top and bottom surfaces of the cut sections because of the free expansion of the surfaces in these regions. von Mises stress attains high values at the mid-thickness of the cut section which is due to the constraints associated with the thermal expansion in this region. The cut edges are free from sideways burning indicating the proper selection of the cutting parameters. SEM micrographs reveal that the kerf surface is free from the striations and cracks. The use of nitrogen as assisting gas suppresses the excessive heat generation due to high temperature oxidation reaction in the cut section. Some locally distributed small dross attachments are observed at the kerf exit, which is attributed to the low drag force generated on the molten flow and the surface tension effect of the molten material at the kerf exit.

Acknowledgments The authors acknowledge the support of King Fahd University of Petroleum and Minerals, Dhahran, Saudi Arabia, for this work.

References

1. Kathuria YP (2003) A preliminary study on laser assisted aluminum foaming. J Mater Sci 38(13):2875–2881
2. Schmid SR, Nebosky PS, Stalcup G (2009) A manufacturing framework for biomimetic porous metals. Trans North Am Manuf Res Inst SME 37:183–188
3. Fujimura T, Norimatsu T, Nakai M, Nagai K, Iwamoto A, Mima K (2007) Laser machining of RF foam by second harmonics of Nd: YAG laser. Fusion Sci Technol 51(4):677–681
4. Campana G, Bertuzzi G, Tani G, Bonaccorsi LM, Proverbio E (2008) Experimental investigation into laser welding of aluminum foam filled steel tubes. MetFoam 2007. In: Proceedings of the 5th international conference on porous metals and metallic foams, pp 453–456
5. Carcel B, Carcel AC, Perez I, Fernandez E, Barreda A, Sampedro J, Ramos JA (2009) Manufacture of metal foam layers by laser metal deposition. In: Proceedings of SPIE—the international society for optical engineering, vol 7131
6. Ocelík V, van Heeswijk V, De Hosson JThM, Csach K (2004) Foam coating on aluminum alloy with laser cladding. J Laser Appl 16(2):79–84
7. Yoshida Y, Yajima Y, Hashidate H, Ogura H, Ueda S (2002) Hole drilling of glass-foam substrates with laser. In: Proceedings of SPIE—the international society for optical engineering, vol 4426. pp 154–157
8. Guglielmotti A, Quadrini F, Squeo EA, Tagliaferri V (2009) Laser bending of aluminum foam sandwich panels. Adv Eng Mater 11(11):902–906
9. Yilbas BS, Akhtar SS, Karatas C (2012) Laser straight cutting of alumina tiles: thermal stress analysis. Int J Adv Manuf Technol 58:1019–1030
10. ABAQUS Theory Manual, Version 6.9 (2009) ABAQUS Inc., Pawtucket, USA
11. Mukarami T, Tsumura T, Ikeda T, Nakajima H, Nakata K (2007) Anisotropic fusion profile and joint strength of lotus-type porous magnesium by laser welding. Mater Sci Eng A 456:278–285
12. Coquard R, Baillis D (2009) Numerical investigation of conductive heat transfer in high-porosity foams. Acta Mater 57:5466–5479
13. Scintilla LD, Tricarico L (2012) Estimating cutting front temperature difference in disk and CO_2 laser beam fusion cutting. Opt Laser Technol 44:1468–1479
14. Swinehart DF (1962) The Beer-Lambert law. J Chem Educ 39:333
15. Guo S, Jun H, Lei L, Yao Z (2009) Numerical analysis of supersonic gas-dynamic characteristic in laser cutting. Opt Lasers Eng 47:103–110
16. Ottoa A, Koch H, Leitz K, Schmidt M (2011) Numerical simulations—A versatile approach for better understanding dynamics in laser material processing. Physics Procedia 12:11–20
17. Deshpande VS, Fleck NA (2000) High strain rate compressive behavior of aluminum alloy foams. Int J Impact Eng 24:277–298
18. Ashby MF, Evans AG, Fleck NA, Gibson LJ, Hutchinson JW, Wadley HNG (2000) Metal foams: a design guide. Butterworth-Heinemann, Oxford

Chapter 4
Micro-Electrical Discharge Machining

Muhammad P. Jahan

Abstract This chapter provides a comprehensive overview on the micro-electrical discharge machining (micro-EDM) process. The physical principle of micro-EDM and differences between macro- and micro-EDM and micro-EDM system components are discussed. A discussion on major operating parameters of micro-EDM and their effect on the performance parameters is presented. A brief discussion on variants of micro-EDM and their industrial applications has been included. The research advances on the areas of micro-EDM and micro-EDM-based compound and hybrid machining processes have been discussed. Finally, a comparative evaluation of micro-EDM to other machining processes has been presented.

4.1 Introduction: History of EDM and Micro-EDM

Electrical discharge machining (EDM) is among the earliest non-traditional manufacturing processes, having an inception more than 60 years ago in a simple die-sinking application. The history of the EDM process dates back to the days of World Wars I and II. Earlier, very few saw the benefits of this process and the popularity of the primitive technology was scarce, as much electrode material was removed as that of the workpiece and the manual feed mechanism led to more arcing than sparking. The process of material removal by controlled erosion through a series of sparks, commonly known as electric discharge machining, was first started in the USSR in the 1940s. Two Soviet husband and wife scientists, Doctors B.R. and N.I. Lazarenko, first applied it to a machine for stock removal [1]. They were convinced that many more improvements could be made to control the feed mechanism and then invented the relaxation circuit. They invented a

M. P. Jahan (✉)
Department of Architectural and Manufacturing Sciences,
Western Kentucky University, Bowling Green, KY 42101, USA
e-mail: muhammad.jahan@wku.edu

J. P. Davim (ed.), *Nontraditional Machining Processes*,
DOI: 10.1007/978-1-4471-5179-1_4, © Springer-Verlag London 2013

simple servo controller too that helped maintain the gap width between the tool and the workpiece. This reduced arcing made EDM machining more profitable. This was the turning point in the history of the EDM process. Initially, EDM was used primarily to remove broken taps and drills from expensive parts. Through the years, the machines have improved drastically—progressing from RC (resistor capacitance or relaxation circuit) power supplies and vacuum tubes to solid-state transistors with nanosecond pulsing, from crude hand-fed electrodes to modern CNC-controlled simultaneous six-axes machining.

The two principle types of EDM processes are the die-sinking EDM and the wire EDM. The die-sinking process was refined as early as in the 1940s with the advent of the pulse generators, planetary and orbital motion techniques, CNC, and the adaptive control mechanism. From the vacuum tubes, to the transistors to the present-day solid-state circuits, not only was it possible to control the pulse on time, but the pause time or the pulse off time could also be controlled. This made the EDM circuit better, accurate, and dependable, and EDM industry began to grow. During the 1960s, the College International pour la Recherche en Productique (CIRP) and International Symposium for ElectroMachining (ISEM) conferences were held for the first time in Czechoslovakia, which proved to be a driving force in the progress of the EDM process. The evolution of the wire EDM in the 1970s was due to powerful generators, new wire tool electrodes, better mechanical concepts improved machine intelligence, and better flushing [2]. Over the years, the speed of wire EDM has gone up 20 times when it was first introduced, and machining costs have decreased by at least 30 % over the years. Surface finish has improved by a factor of 15, while discharge current has gone up more than 10 times higher. Figure 4.1 shows the evolution of EDM research and world market through time [3].

While EDM process has found widespread applications in the industry over last 70 years, true micro-EDM came to existence first in 1968 when Kurafuji and Masuzawa [4] demonstrated the first application of micro-EDM by drilling a minute hole in 50-μm-thick carbide plate. Since then, significant amount of research efforts has been focused on the development of micro-machining

Fig. 4.1 Evolution of EDM research and world market through time [3]

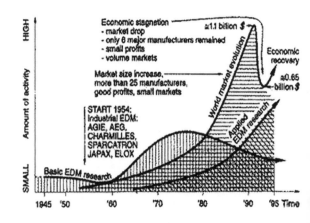

processes. Despite the efforts, initially it had a rather slow industrial acceptance for production processes, until recently when the use of micro-EDM for micro-scale manufacturing became inevitable due to the demand of miniaturization.

4.2 Principle of Micro-EDM

Micro-EDM is the process of machining electrically conductive materials in the form of micro-size craters by using precisely controlled sparks that occur between an electrode and a workpiece in the presence of a dielectric fluid [5]. The basic mechanism of the micro-EDM process is essentially similar to that of the EDM process with the main difference being in the size of the tool used, the power supply of discharge energy, and the resolution of the X-, Y-, and Z-axes movement [6]. The machining process is driven by assigned and controlled gap, voltage, energy, and frequency of discharge. High frequencies (>200 Hz) and small energies (10^{-6}–10^{-7} J) for every discharge (40–100 V) are required to obtain high accuracy and good surface qualities (roughness of about 0.1 µm) [7]. A pulse generator supplies the discharge energy, and a servo system is employed to ensure that the electrode moves at a proper rate to maintain the right spark gap and to retract the electrode if short-circuiting occurs. A dielectric circulation unit with pump, filter, and tank is used to supply the fresh dielectric in the gap and to maintain the proper flushing out of debris.

The sparking phenomena during micro-EDM can be separated into three important phases named as preparation phase for ignition, phase of discharge, and interval phase between discharges [8]. When the gap voltage is applied, an electric field or energy column is created, which gains highest strength once the electrode and surface are closest. The electrical field eventually breaks down the insulating properties of the dielectric fluid. Once the resistivity of the fluid is lowest, a single spark is able to flow through the ionized flux tube and strike the workpiece. The voltage drops as the current is produced and the spark vaporizes anything in contact, including the dielectric fluid, encasing the spark in a sheath of gasses composed of hydrogen, carbon, and various oxides. The area struck by the spark will be vaporized and melted, resulting in a single crater. Due to the heat of spark and contaminates produced from workpiece, the alignment of the ionized particles in the dielectric fluid is disrupted, and thus, the resistivity increases rapidly. Voltage rises as resistivity increases and the current drops, as dielectric can no longer sustain a stable spark. At this point, the current must be switched off, which is done by pulse interval. During the pulse off time, as heat source is eliminated, the sheath of vapor that was around the spark implodes. Its collapse creates a void or vacuum and draws in fresh dielectric fluid to flush away debris and cool the area. Also, the re-ionization happens which provides favorable condition for the next spark. Figure 4.2 shows the principle of the micro-EDM process [9].

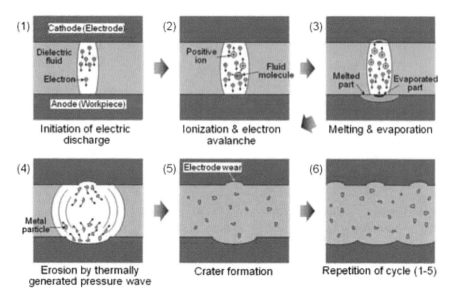

Fig. 4.2 Illustration of the principle of micro-EDM [9]

4.3 Differences Between Macro and Micro-EDM

Even though micro-EDM is based on the same physical principle of spark erosion, it is not merely an adoption of the EDM process for machining at micron level. There are significant differences in the size of the tool used, fabrication method of micro-sized tools, the power supply of discharge energy, movement resolution of machine tools' axes, gap control and flushing techniques, and processing techniques [10–12]. For example, micro-EDM milling, wire electro-discharge grinding (WEDG), and repetitive pattern transfer are commonly employed in and more specific to the micro-EDM process. Some other differences between the macro- and micro-EDMs are listed below:

- The most important difference between micro-EDM and EDM (for both wire EDM and die-sinking EDM) is the dimension of the plasma channel radius that arises during the spark. In conventional EDM, the plasma channel is much smaller than the electrode, but the size is comparable to micro-EDM [13].
- Smaller electrodes (micro-WEDG and micro-BEDG can produce electrodes as small as Ø5 mm and thin wires can be <Ø20 mm) used in the micro-EDM process present a limited heat conduction and low mass to dissipate the spark heat. Excessive spark energy can produce the wire rupture (or electrode burn in die-sinking micro-EDM), being the maximum applicable energy limited by this fact in micro-EDM [13].
- Together with the energy effects, the flushing pressure acting on the electrode varies much in micro-EDM with respect to the conventional EDM process.

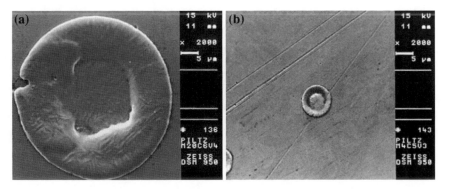

Fig. 4.3 Comparison between crater dimensions in **a** conventional EDM (*left*) and **b** micro-EDM (*right*) [15]

In micro-EDM, the electrode pressure area is smaller, but the electrode stiffness is lower, increasing the risk of electrode breakage or tool deflection. The debris removal is more difficult in micro-EDM because the gap is smaller, the dielectric viscosity is higher, and the pressure drop in micro-volumes is higher [14].

- In the conventional EDM, the higher precision can only be achieved if electrode vibrations and wear are controlled. On the other hand, the precision and accuracy of the final products are much higher in micro-EDM [13].
- For each discharge, the electrode wear in micro-EDM is proportionally higher than conventional EDM. The electrode is softened, depending on the section reduction in the spark energy.
- In micro-EDM, the maximum peak energy must be limited to control the unit removal rate per spark (UR) and use small electrodes and wires. Therefore, the crater sizes in micro-EDM are also much smaller than those in conventional EDM [15]. Figure 4.3 shows the comparison of crater size with conventional EDM and micro-EDM [15].

4.4 Micro-EDM System Components

The EDM and micro-EDM machine are made of several major system components, the central control unit, the servo control, the work tank, the position control, and the dielectric tank. The central control unit manages the operation between the operator and the machine and also those between the different elements of the machine. It also houses the memory units and the power supply or discharging unit. The servo control unit monitors and determines the advancement of the electrode toward the workpiece. The work tank is where the workpiece is clamped, and it is designed to machine when the workpiece is covered with dielectric to prevent fire risks. The position control determines the motion of the

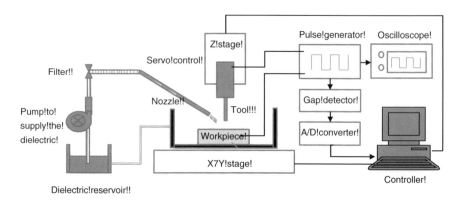

Fig. 4.4 A schematic diagram showing different units of EDM or micro-EDM

tool electrode in the X-, Y-, Z-, and C-axes. The movement is controlled by servo motors, and it also controls the speed of descend of electrode onto the workpiece as determined by the servo parameter. The dielectric tank stores the dielectric. It consists of an electro-valve for filling and emptying and pumps to fill and maintain the dielectric level in the work tank. The rotating unit is added to provide rotation of the electrode. Figure 4.4 shows schematic diagram of the basic units of the EDM or micro-EDM system. A brief description of three major system components of micro-EDM is provided in the following section.

4.4.1 Pulse Generators or Discharging Unit

The controller circuit can have two types of pulse generators—resistance–capacitance (RC) or relaxation-type pulse generator and transistor-type pulse generator. Based on the research by Han et al. [16] and review by Kunieda et al. [17], a description is provided on different types of pulse generators. With growing demands for micro-parts, micro-EDM is becoming increasingly important. However, micro-EDM has poor material removal rate due to the use of conventional pulse generators and feed control systems. In conventional EDM, as mentioned before, two kinds of pulse generators are generally used: relaxation or RC-type pulse generator and transistor-type pulse generator as shown in Fig. 4.5a and b, respectively.

The fabrication of parts smaller than several micrometers requires minimization of the pulse energy supplied into the gap between the workpiece and electrode. This means that finishing by micro-EDM requires pulse duration of several dozen nanoseconds. Since the RC pulse generator can generate such small discharge energy simply by minimizing the capacitance in the circuit, it is widely applied in micro-EDM. However, machining using the RC pulse generator is known to have the following demerits:

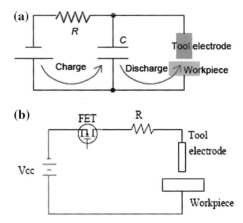

Fig. 4.5 Schematic representation of **a** RC-type and **b** transistor-type pulse generators

1. Extremely low removal rate from its low discharge frequency due to the time needed to charge the capacitor.
2. Uniform surface finish is difficult to obtain because the discharge energy varies depending on the electrical charge stored in the capacitor before dielectric breakdown.
3. Thermal damage on the workpiece when the dielectric strength is not recovered after the previous discharge and the current continues to flow through the same plasma channel in the gap without charging the capacitor.

The transistor-type pulse generator is on the other hand widely used in conventional EDM. Compared with the RC pulse generator, it provides a higher removal rate due to its high discharge frequency because there is no need to charge any capacitor. Moreover, the pulse duration and discharge current can arbitrarily be changed depending on the machining characteristics required. This indicates that the application of the transistor-type pulse generator to micro-EDM can provide dramatic improvements in the removal rate due to the increase in the discharge frequency by more than several dozen times.

In a transistor-type pulse generator, a series of resistances and transistors are connected in parallel between the direct current power supply and the discharge gap. The discharge current increases proportionally to the number of transistors which is switched on at the same time. The FET operates the switching on–off of gate control circuit. In order to generate a single pulse, gap voltage is monitored to detect the occurrence of discharge and after preset discharge duration, the FET is switched off. However, there is a delay in signal transmission from the occurrence of discharge to the switching off of the FET due to the time constants in voltage attenuation circuit, pulse control circuit, and insulating circuit and gate drive circuit for the FET [16].

In a RC- or relaxation-type circuit, discharge pulse duration is dominated by the capacitance of the capacitor and the inductance of the wire connecting the capacitor to the workpiece and the tool [18]. The frequency of discharge (discharge repetition rate) depends on the charging time, which is decided by the

Fig. 4.6 Charge stored in
both stray capacitance and
condenser is discharged

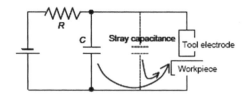

resistor (R) used in the circuit. Therefore, R should not be made very low because arcing phenomenon can occur instead of sparking and a critical resistance is desirable which will prevent arcing. Discharge energy is determined by the used capacitance and by the stray capacitance that exists between the electric feeders, tool electrode holder, and worktable and between the tool electrode and workpiece (Fig. 4.6). This means that the minimum discharge energy per pulse is determined by the stray capacitance. Hence, all the charge stored in the stray capacitance is discharged to the working gap together with the charge stored in the capacitor wired to the circuit. In the final finishing, when minimum discharge energy is necessary, the capacitor is not wired and machining is conducted with the stray capacitance only [18]. It can easily generate pulses with high peak current values and short duration, allowing efficient and accurate material removal, and meanwhile achieving the required surface quality. Finally, pulse conditions with shorter discharge duration and higher peak current provide better surface roughness due to smaller discharge crater [17].

Early EDM equipment used relaxation-type pulse generators with capacitor discharges as shown in Fig. 4.5a. This type of equipment has been used especially where discharge current with high peak values and short duration is needed. With improved capability of power transistors that can handle large currents with high response, the transistor type shown in Fig. 4.5b replaced the relaxation type. However, the relaxation-type pulse generators are still being used in finishing and micromachining because it is difficult to obtain significantly short pulse duration with constant pulse energy using the transistor-type pulse generator. If the transistor type is used, it takes at least several tens of nanoseconds for the discharge current to diminish to zero after detecting the occurrence of discharge because the electric circuit for detecting the occurrence of discharge and the circuit for generating an output signal to switch off the power transistor and the power transistor itself have a certain amount of delay time. Hence, it is difficult to keep the constant discharge duration shorter than several tens of nanoseconds using the transistor-type pulse generator.

4.4.2 Servo Control System or Gap Control Unit

In the micro-EDM process, to ensure stable and efficient machining, the gap condition between the electrode and workpiece must be maintained at the desired state by servo-controlled electrode feed mechanism. The accurate detection of discharge states and stable servo feed control method are the key technologies that are the prerequisite and guarantee of high-efficiency and high-stability processing [19]. A stable gap control system also enables better dimensional accuracy of micro-machined features [18].

During the micro-EDM, the discharging gap between the tool electrode and workpiece is several microns or even less [20] and therefore requires design of special servo controls compared to conventional EDM. The micro-controller is the most important module of a servo control system, which sends control digital signal, translates analog to digital signal by A/D module, and drives the motor by PWM module [20]. To every micro-feeding, shift detection result must be feed-back to the micro-controller, to adjust the output shift in order for precise control of servo control system. The servo control can be operated based on different algorithms or principles like predicting the gap distance and offsetting tool position, ignition delay time, average gap voltage, the average delay time, etc. [18].

Ignition delay time (t_d) is an important indicator of the isolation condition of discharge gap. Larger gap width causes longer ignition delays, resulting in a higher average voltage. Tool feed speed increases when the measured average gap voltage is higher than the preset servo reference voltage and vice versa [21]. On the contrary, the feed speed decreases or the electrode is retracted when the average gap voltage is lower than the servo reference voltage, which is the case for smaller gap widths resulting in a smaller ignition delay. Thus, short circuits caused by debris particles and humps of discharge craters can be avoided. Also, quick changes in the working surface area, when tool electrode shapes are complicated, do not result in hazardous machining. In some cases, the average ignition t_d is used in place of the average gap voltage to monitor the gap width [17]. In other attempts, gap-monitoring circuits were developed to identify the states and ratios of gap open, normal discharge, transient arcing, harmful arcing, and short circuit [22]. These ratios were used as input parameters for online EDM control based on various control strategies. The servo feed control shown in Fig. 4.7 keeps the working gap at a proper width [17].

4.4.3 Dielectric Circulation System or Flushing Unit

The dielectric circulation system is the integral part of the micro-EDM system. The dielectric system consists of dielectric fluid, dielectric reservoir, pump to supply the dielectric fluid into the work tank, filter to remove the debris particles from the dielectric and to ensure re-circulation of fresh dielectric to the machining zone, pipe and nozzles to supply the dielectric into the gap between the workpiece and the electrode, and flushing pressure control valve to maintain/control the flushing pressure of the dielectric. Dielectric fluid is the most important component of the flushing unit, provides a known electrical barrier between the electrode and workpiece, and acts as a means for the removal of spark debris from the spark gap.

Fig. 4.7 Principle of servo
control system based on
ignition delay [17]

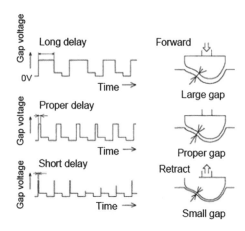

The major functions of the dielectric circulation system or flushing unit are as
follows:

- To distribute the dielectric flow through the spark gap to remove gaseous and
  solid debris generated during micro-EDM.
- To introduce fresh and clean dielectric fluid to the machining region.
- To flush away the chips or metal particles generated in the spark gap.
- To maintain the dielectric temperature well below its flash point.
- To act as a cooler for cooling the electrode and workpiece.

4.5 Major Parameters of Micro-EDM

4.5.1 Operating Parameters

In micro-EDM, identifying the major parameters is the first step before proceeding
to find the optimum parameters. From literatures concerning EDM and micro-
EDM, the following parameters are identified as the major ones:

- Gap voltage;
- Peak current;
- Pulse on time (T_{on});
- Pulse off time (T_{off}).

4.5.1.1 Gap Voltage

It is the voltage applied between the tool and the workpiece. The applied voltage
determines the total energy of the spark. If the voltage is high, the erosion rate
increases and the higher machining rate is achieved. But at the same time, higher

voltage will also contribute to poor surface roughness. In order to achieve higher machining rate, higher voltage should be used which might result in higher tool wear. Therefore, for micro-EDM, a very moderate value of voltage needs to be used.

4.5.1.2   Peak Current

This is another very important parameter that determines almost all the major machining characteristics such as machining rate, surface roughness. During machining, the current level fluctuates. The term 'peak current' is often used to indicate the highest current during the machining. The higher the peak current setting is, the larger is the discharge energy. From experimental evidences of previous research work, it seems that sensitivity of the peak current setting on the cutting performance is stronger than that of the pulse on time. When the peak current setting is too high, it may lead to higher tool wear as well.

4.5.1.3   Pulse on Time

This is the duration of time (usually expressed in μs); the current is allowed to flow per cycle. Material removal rate is directly proportional to the amount of energy applied during this pulse on time. This energy is controlled by the peak current and the length of the pulse on time. The main EDM operation is effectively done during this pulse on time. It is the 'work' part of the spark cycle. Current flows and work is done only during this time. Material removal is directly proportional to the amount of energy applied during this time. With longer period of spark duration, the resulting craters will be broader and deeper; therefore, the surface finish will be rougher. Shorter spark duration, on the other hand, helps to obtain fine surface finish.

4.5.1.4   Pulse off Time

This is the duration of time (μs) between two successive sparks when the discharge is turned off. Pulse off time is the duration of the rest or pause required for re-ionization of the dielectric. This time allows the molten material to solidify and to be washed out of the spark gap. If the pulse off time is too short, it will cause sparks to be unstable, and then, more short-circuiting will occur. When the pulse off time is shorter, the number of discharges with a given period becomes more. This results in higher machining speed, but the surface quality becomes poor because of a larger number of discharges. On the other hand, a higher pulse off time results in higher machining time. Although larger pulse off time slows down the process, it can provide stability required to successfully EDM a given application. When the pulse off time is insufficient as compared to on time, it will cause erratic cycling and retraction of the advancing servomotors, slowing down the operation.

4.5.1.5 Capacitance

For RC-type pulse generators, the main parameters are voltage, resistance, and capacitance. The voltage parameter can be explained the same as in transistor-type pulse generators. The different element here is the capacitance parameter. The capacitor in the circuit 'charges' during part of the cycle and then 'discharges' during the machining period. So this parameter is directly related to the discharge energy. Higher value of capacitance thus means more energy per cycle and vice versa. The higher the discharge energy, the faster the machining rate. However, very high settings of capacitor can result in poor surface finish and machining instability.

4.5.1.6 Resistance

The resistance is one of the three important parameters for micro-EDM using the RC-type pulse generator. The change of resistance in effect changes the amount of current applied for micro-EDM. The applied energy is thus a function of the resistance. If the resistance is increased, the amount of current flowing through the circuit is reduced, thus reducing the total discharge energy. Therefore, for faster machining, lower setting of resistance is desired. However, low resistance setting can result in instable machining. Hence, optimum selection of resistance is important that will provide moderately higher material removal rate, desired surface finish, and improved machining stability.

The actual voltage and current waveforms of the transistor- and RC-type pulse generators are shown in Fig. 4.8 [23].

4.5.2 Performance Parameters

The major performance measures or machining characteristics that are generally studied in the micro-EDM are as follows:

- Spark gap;
- Material removal rate (MRR);
- Surface quality or surface integrity;
- Tool wear ratio.

In micro-EDM, the phenomena relating to the parameters are complex and mostly stochastic in nature. Thus, it puts forward the challenges in the understanding of the effects and interaction of the parameters. In order to get desired micro-machining result in micro-EDM, the operating parameters should be optimized to get the following machining characteristics:

- Minimization of spark gap;
- Increase in material removal rate, that is, reduction in machining time;

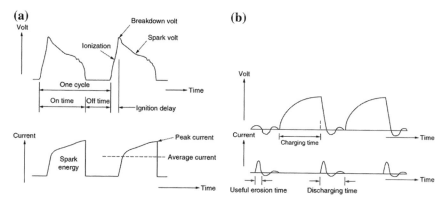

Fig. 4.8 The actual voltage and current characteristics curve for **a** transistor-type and **b** RC-type pulse generators [23]

- Improvement in surface quality;
- Reduction in tool wear.

4.5.2.1 Spark Gap

For EDM or micro-EDM, there is always be a small space, known as the spark gap, between the electrode and the workpiece. This spark gap affects the ability to achieve good dimensional accuracy and good finishes. The lower and consistent the size of the gap is, the more predictable will be the resulting dimension. The machining accuracy depends on the minimum spark gap possible. In order to achieve micro-features, the spark gap should be very small. Thus, for micro-EDM, it is a major challenge to reduce the gap width as much as possible. The spark gap and surface roughness are mainly influenced by pulse on time. But it was found out that current and the applied energy influence the spark gap. Therefore, the main parameters affecting spark gap were identified as open-circuit voltage, peak current, and pulse on time [24].

The dimensional accuracy is very important in cutting micro-parts. For micro-EDM, it is of practical need that the spark gap and hence the dimension of the machined part should be predictable and under control. Depending on different machining conditions, this gap may vary. In order to have dimensional accuracy, there is a need to know how to control the spark gap.

4.5.2.2 Material Removal Rate

Material removal rate (MRR) in micro-EDM is defined as the amount of material that is removed per unit time. It is an indication of how fast or slow the machining rate is. Since machining rate is related to the economic aspect, often it is of high preference objective to achieve. Thus, a parameter that leads to higher MRR is important for production. This is more so in micro-EDM, as this is usually a slow process. At the same time, higher machining productivity must also be achieved with a desired accuracy and surface finish. The MRR is usually calculated using the following equation [25]:

$$\text{Material removal rate (MRR)} = \frac{\text{Volume of material removed from part}}{\text{Time of machining}}$$

An increase in peak current leads to the increase in the volumetric metal removal rate [26]. These results have been attributed to the fact that an increase in peak current leads to the increase in the rate of heat energy and hence the rate of melting and evaporation. However, after a certain value, due to arcing, the machining efficiency and subsequently MRR decrease. When the flushing pressure increases, the tendency of arcing decreases and the MRR increases. MRR generally increases with the increase in the duty factor, which is defined as the ratio of pulse on time to total pulse on and off time. At higher value of duty factor, same heating temperature is applied for a longer time. This causes an increase in the evaporation rate and gas bubbles number, which while exploding cause the removal of bigger volume of molten metal.

4.5.2.3 Surface Roughness

During each electrical discharge, intense heat is generated that causes local melting or even evaporation of the workpiece material. With each discharge, a crater is formed on the workpiece. Some of the molten material produced by the discharge is carried away by the dielectric fluid circulation, while the remaining melt re-solidifies to form an undulating terrain around the machined surface. To improve the micro-EDMed surface integrity, the size of craters needs to be small [27]. It was found that surface roughness increased when the pulse on time and open-circuit voltage were increased. Because of greater discharge energy, the surface roughness is affected by pulse on time and open-circuit voltage [27]. Again, depending on the nature of the work material, the surface roughness varies. The surface roughness shows slightly decreasing trend with increasing flushing pressure. The cutting performance with increasing dielectric fluid pressure improves because the particles in the machining gap are evacuated more efficiently. It is also demonstrated that the surface roughness slightly increases with the increase in peak current value up to certain level and then vigorously increases with any increase in peak current [26]. It can be explained by the fact that increase in peak current causes an increase in discharge heat energy at the point where the discharge takes place. The overheated pool of molten metals evaporate forming gas bubbles that explode when the discharge ceases. This takes molten metals away and forms crater on the surface. Successive discharges thus result in worse surface roughness.

4.5.2.4 Tool Wear Ratio

The tool wear ratio is defined as the volume of metal lost from the tool divided by the volume of metal removed from the workpiece. High tool wear rates result in inaccurate machining and add considerably to the expense since the tool electrode itself must be first accurately machined. Tool wear ratio is calculated by the following equation [25]:

$$\text{Total wear ratio} = \frac{\text{Volume of material removed from electrode}}{\text{Volume of material removed from part}}$$

One of the problems in EDM/micro-EDM is that material not only gets removed from the workpiece, but also gets removed from the tool while machining. Because of this, the desired shape often is not achieved due to the lack of accuracy in the deformed geometry of the tool. To reduce the influence of the electrode wear, it is necessary either to feed electrode larger than the workpiece thickness in the case of making through holes or to prepare several electrodes for roughing and finishing in the present state of technology [28]. Tool life is an important concern in micro-EDM. The low energy range is becoming important when the EDM process is used in the micro-field. However, little is known of the electrode wear in low energy range applied in micro-EDM. Tsai and Masuzawa [28] experimentally investigated the electrode wear of different materials and found the followings:

- The volumetric wear ratio of the electrode becomes small for the electrode material with high boiling point, high melting point, and high thermal conductivity. This tendency is independent of the workpiece materials.
- Corner wear of electrode relates to diffusion of heat. The corner rounding is more obvious when the thermal conductivity of the electrode is low.
- The boiling point of the electrode material plays an important role in wear mechanism of micro-EDM, since high surface temperature and high energy density correspond to small discharge spot.

Therefore, the tool wear characteristics are primarily associated with material properties. In addition, the wear of the electrode is related to such factors as the distribution of discharge power between both electrodes and the thermodynamic constants of materials [29]. In addition, the tool wear ratio increases with the increase in gap voltage, peak current, and pulse duration. The higher values of pulse interval are better in terms of tool wear ratio; however, very higher settings of pulse interval can again increase the tool wear ratio. Therefore, suitable selection of operating parameters is a key to determine optimum tool wear ratio in micro-EDM.

In the case of drilling micro-holes using micro-EDM, tool wear causes a taper in the achieved holes. Due to the uneven tool wear of the electrode, the diameter of the hole on the top surface will be different from that at the bottom surface. The taper may also be associated with the corner wear of the tool electrode. The value of the taper in micro-hole machining is calculated by the following equation [30]:

$$\text{Taper} = \frac{\text{Diameter of hole on top surface} - \text{Diameter of hole on bottom surface}}{2 \times \text{Thickness of workpiece, h}}$$

4.6 Variants of Micro-EDM

4.6.1 Die-Sinking Micro-EDM

Die-sinking micro-EDM is the earliest and most common type of micro-EDM process. In die-sinking micro-EDM, an electrode with desired micro-features is employed on the workpiece to produce corresponding mirror images. The tool electrode has the complementary form of the finished workpiece and literally sinks into the workpiece. The servo controller monitors the gap conditions (voltage and current) and synchronously controls the different axes to machine the mirror image of the tool. Figure 4.9 shows the schematic representation of the die-sinking micro-EDM [31].

4.6.2 Micro-Wire Electro-Discharge Machining

In micro-wire electro-discharge machining (micro-WEDM), a continuously traveling micro-wire is used to cut through a conductive workpiece according to the programmed path. The basic mechanism of micro-WEDM is same as micro-EDM and the material removed as a result of series of electric sparks between the

Fig. 4.9 Schematic showing the principle of die-sinking micro-EDM [31]

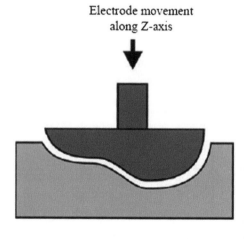

Electrode movement along Z-axis

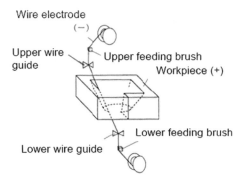

Fig. 4.10 Schematic showing the principle of micro-WEDM [17]

workpiece and wire electrode. Since wire orientation can be changed by controlling the horizontal position of the upper wire guide relative to the lower guide, all types of ruled surfaces can be cut by wire-WEDM process. Figure 4.10 represents the schematic representation of micro-WEDM [17].

4.6.3 Dry and Near-Dry Micro-EDMs

Dry or near-dry micro-EDM is a recent variant of micro-EDM that has been found to decrease the pollution caused by the use of liquid dielectric, which leads to production of vapors during machining and adds cost to manage the waste, thus suitable for green manufacturing. The dry or near-dry micro-EDM can be die-sinking type or milling type. The dry micro-EDM uses oxygen or air as the dielectric medium for machining [32]. On the other hand, the near-dry micro-EDM uses liquid–gas mixture or mist as the dielectric. The mist dielectric can be mixture of gas medium (air/nitrogen) and liquid (water/kerosene) [33]. In this process, usually a thin walled tubular electrode is used through which high-pressure gas or air is supplied to the machining zone. The role of the gas is to act as dielectric, remove the debris from the gap, and cool the inter electrode gap. Although the dry and near-dry micro-EDMs have been found to be a variant of die-sinking or milling micro-EDM with gas medium of dielectric, there are significant differences in plasma characterization, gap control, or even the material removal mechanism [34, 35]. The principle of dry micro-EDM [35] and dry micro-EDM milling [33] is presented in Fig. 4.11.

4.6.4 Micro-EDM Drilling

In micro-EDM drilling, micro-electrodes are used to 'drill' micro-holes in the workpiece. However, the problem with deep small-hole drilling by micro-EDM is that forming and clamping the long electrode are difficult. Therefore, the

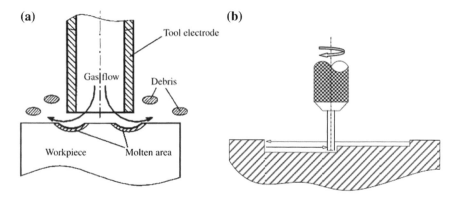

Fig. 4.11 a Principle of the dry micro-EDM using high-pressure gas/oxygen flow through the tube electrode [35]. b Principle of dry milling micro-EDM [33]

micro-electrodes are used to fabricate in situ with high accuracy using different micro-electro-discharge grinding (micro-EDG) processes [discussed later in Sect. 4.6.5]. This also solves the clamping problem, because the electrode is clamped before the machining process and never re-clamped until the hole is machined. Figure 4.12 shows the consecutive process of on-machine electrode fabrication and measurement of micro-electrode and micro-EDM drilling with fabricated electrode [36].

4.6.5 Micro-Electro-Discharge Grinding Process

4.6.5.1 Block Micro-EDG

One of the most commonly used micro-EDM variants is the micro-EDG. During micro-EDM, to fabricate micro-electrode on-machine from an electrode thicker than the required one, micro-EDG process with a sacrificial electrode is used. Block micro-EDG is a simple process requiring a precise sacrificial rectangular block with high wear resistance (WC was used mostly due to its high resistance to wear) and a commercially available electrode. However, one important thing is the alignment of block respective to the electrode. It is very important that sacrificial block should be aligned properly (within an accuracy of ±2 μm) in order to avoid electrodes being more taper, thus reducing dimensional accuracy. It has been found that due to wear of the sacrificial block also, the diameter of the fabricated electrode is sometimes difficult to predict. Therefore, an on-machine camera with measuring unit is installed to measure the dimension in situ. In this method, the block is used as a cutting electrode and a cylindrical rod is used as the workpiece. The micro-electrode that needs to be machined is fed against the conductive block. The machining is carried out at different conditions by applying a controlled

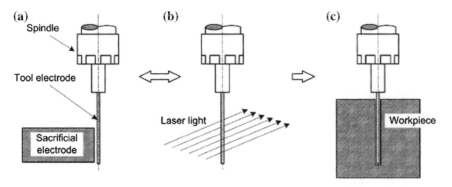

Fig. 4.12 Micro-EDM drilling process; **a** on-machine electrode fabrication by block electro-discharge grinding (BEDG), **b** on-machine measurement by laser, and **c** drilling of high-aspect-ratio micro-holes [36]

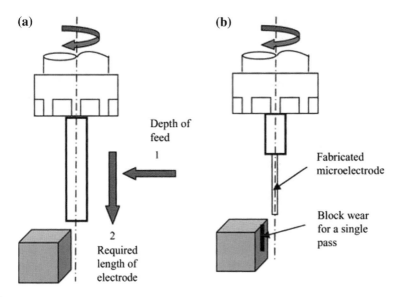

Fig. 4.13 Schematic showing the principle of block micro-EDG process [37]

electric spark and by forcing the dielectric medium to flow through the spark gap between the block and the rod. Figure 4.13 shows the principle of block micro-EDG process [37].

4.6.5.2 Micro-Wire Electro-Discharge Grinding

Micro-wire electro-discharge grinding (micro-WEDG) is a micro-fabrication process that uses electrical discharges in a dielectric fluid to erode material from

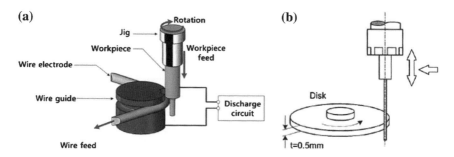

Fig. 4.14 Schematic showing the principle of **a** micro-wire electro-discharge grinding (micro-WEDG) [39] and **b** micro-EDG using rotating disk as sacrificial electrode [36]

conductive workpieces and produce micro-shafts. The discharges occur across a small gap (~ 2 μm) filled with dielectric oil. The workpiece is held vertically in a mandrel that rotates at several thousand rpm, and its position is slowly fed in the Z direction. The wire is supported on a wire guide, and its position is controlled in the X- and Y-directions. Each electrical discharge erodes material from the workpiece and the wire. To prevent discharges from worn regions of the wire, the wire travels at a fixed traveling speed and is fed from a reel and take-up system. The micro-WEDG was first invented by Masuzawa et al. [38], described as a means to manufacture micro-cylindrical electrodes and have made a great change in the miniaturization, as micro-WEDG was found to fabricate very thin micro-EDM electrode with very higher aspect ratio.

In another similar process, rotating disk can replace the moving wire and reduce the chance of wire breakage during the machining process. However, the use of a rotating disk involves a rather complicated setup, although it provides good shape accuracy [36]. Figure 4.14 shows the schematic representation of micro-WEDG [39] and EDG using sacrificial rotating disk [36].

4.6.6 Planetary or Orbital Micro-EDM

The planetary or orbital micro-EDM using the movement of the electrode was found to be very useful particularly in drilling of blind micro-holes as flushing is more difficult for a thin electrode [40]. A common problem during the die-sinking micro-EDM or micro-EDM drilling is the debris accumulation, which becomes worst during machining high-aspect-ratio micro-structures. Adding a relative motion between the electrode and workpiece, other than the electrode feeding motion, produces a wide clearance between them for fluid circulation and then reduces debris concentration, resulting in a high MRR, low electrode wear ratio, and higher machining accuracy. This results in lesser wear of the bottom edges of the tool and therefore minimizes undesirable tapering and waviness at the bottom surface of the blind micro-hole [41]. The tool path depends on the complexity of

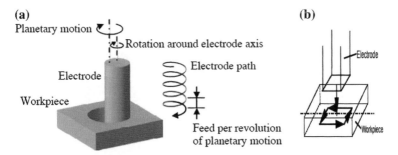

Fig. 4.15 **a** Schematic representation of planetary micro-EDM for circular micro-hole [40]. **b** Planetary movement for non-circular micro-hole [41]

the feature to be machined. Besides improving the MRR and decreasing tool electrode wear, another advantage of orbital micro-EDM in drilling of micro-holes is that it can fabricate micro-holes of different diameters with same electrode size by changing the orbit radius [40]. Figure 4.15a and b represents the schematic of planetary micro-EDM for circular and square micro-holes, respectively [40, 41].

4.6.7 Reverse Micro-EDM

The reverse micro-EDM is being accomplished by reversing the polarity so that the material erodes from the larger diameter rod. The reverse micro-EDM comprises different steps including fabrication of single micro-electrode using micro-EDG process, fabrication of arrays of micro-holes which will act as negative electrode during reverse EDM, and finally fabrication of multiple electrodes using reverse micro-EDM. The polarity of the electrode and sacrificial workpiece is interchanged during reverse micro-EDM, so that the electrode can be extruded among the arrays of micro-holes. In this process, the electrode is considered as workpiece and is fed down to the holes of the metal plate to be machined by electrical discharge, which occurs within machining gap. However, the regions, which correspond to the holes, are not machined. Finally, micro-electrodes are machined as many as the holes. The principle of reverse micro-EDM is presented schematically in Fig. 4.16 [42].

4.7 Application Examples of Micro-EDM Variants

In recent years, micro-EDM has been found to be an important process for the fabrication of micro-components and micro-parts for different industrial applications. In the following section, some of the important industrial applications of micro-EDM variants are provided in brief:

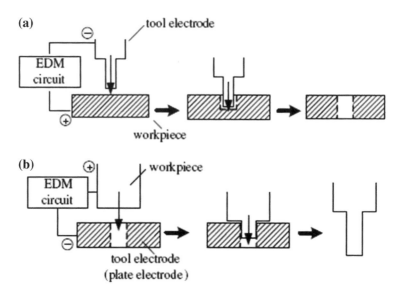

Fig. 4.16 Principle of reverse micro-EDM; **a** fabrication of micro-hole using normal micro-EDM and **b** fabrication of micro-electrode using reverse micro-EDM [42]

- *Automotive Nozzles*: The automotive nozzles are very high-aspect-ratio micro-holes that are machined in very difficult-to-cut materials. The conventional machining processes are difficult to apply for machining of such high-aspect-ratio holes in hard materials. Micro-EDM drilling is mainly used for the machining of high-aspect-ratio automotive nozzles.
- *Spinnerets*: Spinnerets are the special tools used for synthetic fibers. The spinnerets are difficult to fabricate using other machining processes, whereas can be fabricated easily using the micro-EDM process. The spinnerets can be fabricated by several varieties of micro-EDM like micro-WEDG, micro-EDM milling, or die-sinking micro-EDM (Fig. 4.17).
- *Micro-molds*: Micro-molds are the most important micro-parts for the mass fabrication of plastic and electronic components in MEMS industries. Micro-EDM milling is mainly used to fabricate those micro-molds using high-aspect-ratio micro-tool of as small as tens of micron. First, the edges of the micro-molds are machined using micro-WEDM, and then, the slots or cavities could be machined by micro-EDM milling.
- *Fiber-optics and Optics*: Micro-EDM is capable of fabricating micro-features for fiber-optics applications. The figure shows spinnerets tools for synthetic fibers fabricated by micro-EDM process (Fig. 4.18).
- *Aerospace*: The applications of micro-EDM in aerospace could be vast. The micro-wire EDM can be used to cut the difficult-to-cut aerospace materials like titanium and inconel alloys to any desired shape. In addition, micro-EDM drilling is used extensively for fabricating micro-holes in difficult-to-cut

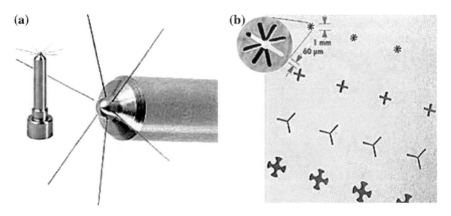

Fig. 4.17 **a** Fuel injector for diesel engine in automobile [high-aspect-ratio micro-hole of diameter down to 0.1 mm (100 μ) machined by micro-EDM]. **b** Spinnerets tools for synthetic fibers (courtesy of Sarix Micro-EDM machine) [43]

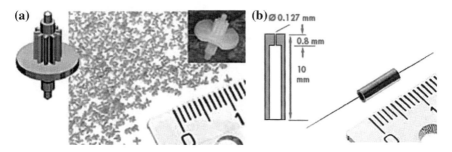

Fig. 4.18 **a** Plastic gear for watches [metallic micro-molds (*left*) cut by the micro-WEDM process; plastic micro-molds can be fabricated from the micro-die fabricated by milling micro-EDM]. **b** Spinnerets tools for synthetic fibers (courtesy of Sarix Micro-EDM machine) [43]

materials, which act as cooling holes. Micro-EDM drilling can also fabricate high-aspect-ratio holes for fuel injection.

- *Medical Instruments*: Micro-EDM is also used for the production of critically shaped medical instruments in difficult-to-cut materials. Both the micro-WEDM and micro-EDM drillings are used extensively for cutting and fabricating holes in different instruments (Fig. 4.19).
- *Micro-tools*: This is one of the most important applications of micro-EDM. The advantage of micro-EDM is that the tools that will be used in micro-EDM can be fabricated on-machine using the same process, maybe by using different varieties of micro-EDM. Different varieties of micro-EDG, that is, block micro-EDG, micro-WEDG, etc. are used to machine those micro-tools. Those micro-tools can be used further for micro-EDM or can be used as cutting tools for some other micro-machining processes.

(a)   (b)

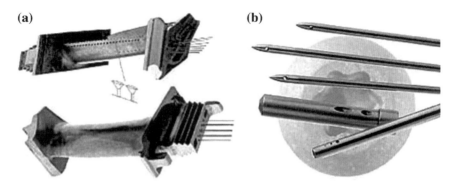

Fig. 4.19 a Cooling holes in aerospace engine components [micro-EDM drilling is used to machine those holes in difficult-to-cut aerospace materials]. b Medical instruments (courtesy of Sarix Micro-EDM machine) [43]

• *Micro-electronics*: Micro-electronics is another major application area for micro-EDM. The micro-EDM process is either used for the fabrication of micro-molds for plastic micro-electronics parts or can be directly used to fabricate different micro-structures in conductive and semi-conductive materials. The micro-WEDM and micro-EDM drilling processes are used extensively for cutting complex shapes and drilling holes, respectively, in micro-electronic parts and components (Fig. 4.20).

(a)   (b)

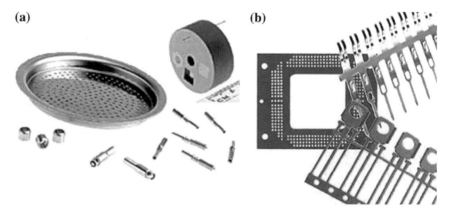

Fig. 4.20 a Micro-stamping tool fabricated by micro-EDM process. b Micro-electronics components fabricated by the micro-EDM process [the circular and non-circular holes are fabricated by micro-EDM drilling, and the edges are machined through micro-WEDM process] (courtesy of Sarix Micro-EDM machine) [43]

4.8 Research Advances and Innovative Applications: Micro-EDM-Based Compound and Hybrid Processes

In recent years, one of the major focuses in the area of advanced research on micro-EDM is the development of micro-EDM-based compound and hybrid processes. The objective is to include the benefit of more than one process and to solve the issues faced by micro-EDM as a single process. Compound machining is defined as the combination of two different machining processes in a single setup applied one after another. On the other hand, the hybrid machining process is defined as the integrated application or combination of different physically active principles in a single process. In the following section, brief discussion is presented on the different micro-EDM-based compound and hybrid processes and their innovative applications.

4.8.1 Micro-EDM-Based Compound Processes

4.8.1.1 Micro-EDM and Micro-Turning

This hybrid process is the combination of micro-EDM and micro-turning processes in a single setup. First, a commercially available PCD tool is modified by micro-EDG process to reduce the nose radius of the cutting tool, thus minimizing the force component that causes the shaft deflection during micro-turning. Commercially available PCD inserts, designed for light finishing cut, have a relatively large tool nose radius, for example 100 µm. This tool nose resolves the cutting force on the shaft into two components, namely F_x and F_y, as can be seen in Fig. 4.21a. The F_y component of the cutting force does the actual cutting, while the F_x component causes deflection of the micro-shaft. A commercially available PCD insert is modified using the micro-EDG process to achieve a very sharp cutting edge, so as to reduce the F_x component of the cutting force significantly, which is illustrated in second image of Fig. 4.21a. This modification of the cutting tool makes it possible to achieve a straight shaft of much smaller diameter. Dual cutter setup is arranged for micro-turning, one with round nose for initial turning up to 100 µm and then sharp tool for final cut up to 20 µm. After the micro-turning process, the fabricated micro-electrodes are used in machining of small- and higher-aspect-ratio micro-holes by micro-EDM on the same machine (Fig. 4.21b). Therefore, this compound process is in fact combination of three processes: micro-EDG, micro-turning, and micro-EDM, one after another in the same machine. Figure 4.21a and b shows the schematic representation of micro-EDM–micro-turning compound process. The fabricated micro-electrode and machined hole using the micro-electrode are presented in Fig. 4.21c and d [44].

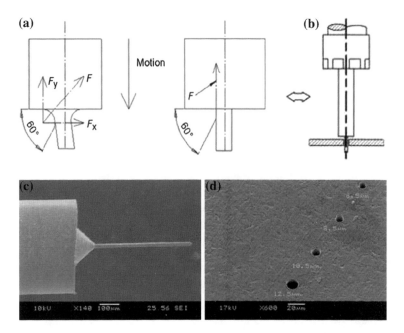

Fig. 4.21 a Modification of the conventional cutting tool using the micro-EDG (variant of micro-EDM) process. **b** Micro-EDM drilling with the tools fabricated by micro-turning with modified cutting tool. **c** 19-μm graphite electrode of 0.5 mm length fabricated by μ-EDM and μ-turning compound process. **d** Fabricated micro-holes with the micro-electrode obtained by the micro-EDM–micro-turning compound process [44]

4.8.1.2 Micro-EDM and Micro-Grinding

In this process, a PCD tool is fabricated on-machine in a desired shape using the micro-EDG process. Thereafter, the fabricated micro-electrode is used for grinding of brittle and hard glass materials with the help of diamond particles that extrude from the matrix metal and act as cutter. When the PCD tool is fabricated by micro-EDG, the binder materials (usually nickel or WC) are removed, as they are conductive, thus protruding the diamond particles, which are non-conductive. Polycrystalline diamond (PCD) with cobalt binder, which can be shaped with micro-EDG, is emerging as a tool material for micro-grinding of hard and brittle materials. The cobalt binder provides an electrically conductive network that can be removed with EDM [45]. The diamond cutting edges are exposed as the discharges erode away the cobalt binder. In addition to micro-grinding, reaming of micro-holes, grinding of micro-slots, and machining of V-grooves with fabricated PCD tool have been reported [46]. Figure 4.22 shows the different steps of the micro-EDM–micro-grinding compound process with machining example [44].

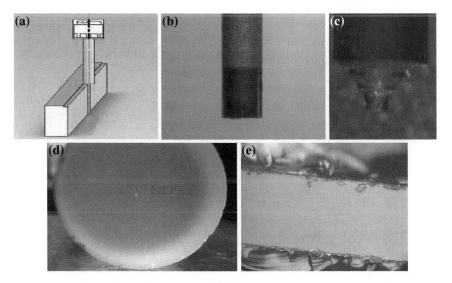

Fig. 4.22 **a** Schematic diagram showing block micro-EDG process (variant of micro-EDM) used to fabricate micro-grinding tool and PCD tool before micro-EDG process. **b** Fabrication of micro-grinding tool with micro-EDG process. **c** Micro-channels on glass machined by micro-grinding process with fabricated PCD tool. **d** Surface finish the micro-channel in glass [44]

4.8.1.3 Micro-EDM and Micro-Milling

This compound process is the combination of micro-WEDG process and micro-milling. In this process, first the micro-milling cutting tool is fabricated out of difficult-to-cut tool materials with the help of micro-WEDG process. After that, the fabricated cutting tool is used for conventional micro-milling operation in the same setup. The use of the micro-WEDG for the production of milling tools has several advantages. The geometry can be changed quite easily, and the potential of scaling down the size of the milling tools is very high [45, 47]. In comparison with other contactless machining technologies, micro-EDM has an acceptable machining time and the resulting costs for the machining are tolerable in comparison with machining with ion beam. An advantage of using the micro-EDM for the milling process is the prevention of inaccuracy by re-chucking processes [47]. Figure 4.23a shows the schematic diagram of the micro-WEDG process of fabricating milling cutting tool. Figure 4.23b and c shows the fabricated cutting tool and machined slot, respectively [47].

4.8.1.4 Micro-EDM and Micro-ECM

This compound process is the combination of micro-EDM and micro-ECM processes in a single setup. The objective is to improve the surface finish generated by micro-EDM using the micro-ECM as a post-processing step. The surface

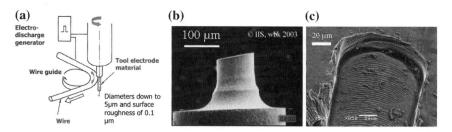

Fig. 4.23 a Schematic diagram of micro-WEDG process used to fabricate micro-milling tools.
b Micro-milling tool fabricated by micro-WEDG. c Micro-slot machined by micro-milling using
fabricated micro-tool [47]

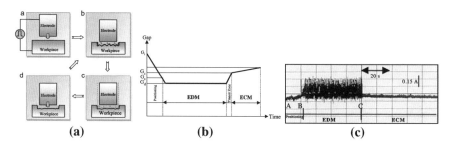

Fig. 4.24 a Schematic diagram showing micro-EDM and micro-ECM compound or hybrid
processes [49]. b The variation in gap distance with time for micro-EDM and ECM in the
compound process. c The change in current due to discharge and dissolution in two stages of
EDM and ECM. [48]

machined by micro-EDM is relatively rough, especially at the micro-level due to
micro-craters and cracks produced by the micro-discharges. Hence, the process
consisting of micro-EDM followed by micro-ECM can be a suitable solution to
improve the machined surface [48]. The de-ionized water used in the micro-EDM
process can serve as an electrolyte solution for micro-ECM under low current
density conditions between electrode and workpiece [49]. The surface after
applying micro-ECM becomes much smoother as well as peak-to-valley distances
of craters (R_{max}) reduced significantly compared to that of micro-EDM. Micro-
ECM can also be applied for finishing the slot machined by micro-EDM milling.
In addition to sequential micro-EDM and micro-ECM processes, combined or
concurrent micro-EDM and micro-ECM processes have also been reported [49,
50], which could be considered as hybrid process. The difference is in hybrid
process, and the discharging of dissolution takes place in the same cycle during
machining, thus applying micro-EDM and micro-ECM concurrently. Figure 4.24
shows the steps of the micro-EDM–micro-ECM compound processes, the change
in gap and current during the EDM and ECM processes of the compound system.
The improvements in surface finish both for compound and for hybrid processes
are presented in Fig. 4.25.

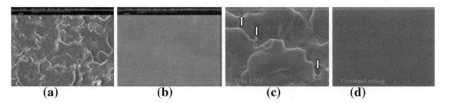

Fig. 4.25 **a** Surface generated by micro-EDM. **b** Surface generated by micro-ECM followed by micro-EDM (compound process). **c** Surface generated by micro-EDM milling process. **d** Surface generated by micro-EDM and micro-ECM combined milling process [50]

4.8.1.5 LIGA and Micro-EDM

This process is used to fabricate high-aspect-ratio micro-electrodes using the LIGA (lithographie, galvanoformung, abformung, in English (X-ray) lithography, electro-plating, and molding) and then applies those micro-electrodes for micro-EDM. The LIGA process uses X-ray lithography to form high-aspect-ratio molds for electro-plated structures. On the other hand, micro-EDM produces 3D micro-structures in any electrically conductive materials. In this compound process, first an array of negative-type electrodes with gear pattern was fabricated in nickel using the LIGA process. The steps of the LIGA process are presented in Fig. 4.26a. After that, a positive-type patterned structure is produced by feeding WC–Co workpiece into one of the electrodes with discharging. Figure 4.26a shows the step-by-step process of fabricating micro-electrode using LIGA process and applying that micro-electrode in micro-EDM [51]. Figure 4.26b–d represents the machining example of this compound process [51].

4.8.1.6 Laser and Micro-EDM

In this compound process, the laser micro-machining is used with the micro-EDM. The laser can be used either for the machining or for welding purposes. A machine, which uses a laser welding system integrated with a micro-EDM process, has been reported [52–54]. The micro-EDM process fabricates the assembled parts, and the Nd-YAG laser performs micro-joining to produce the assembly, so the whole processes from micro-fabrication to micro-assembly can be completed on the same system. This can overcome the problem of small dimensional assembly, and the system, having precision stages ultra-precision motion control technology, can control the positional accuracy precisely between the assembled parts. Figure 4.27a shows the steps of the laser–micro-EDM combined process [53]. The pin-plate micro-assembly done by the combined process is presented in Fig. 4.27b–c [52].

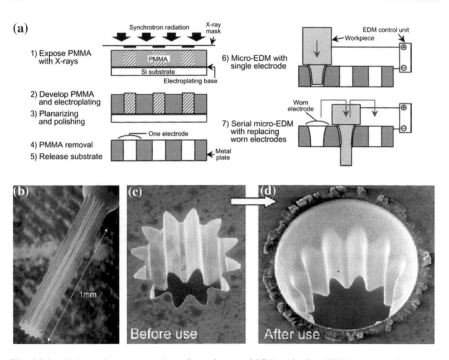

Fig. 4.26 **a** Schematic representation of step-by-step LIGA and micro-EDM compound process. **b** High-aspect-ratio WC–Co micro-structures produced by micro-EDM. **c** Initial negative-type electrode before micro-EDM. **c** Negative-type electrode after micro-EDM [51]

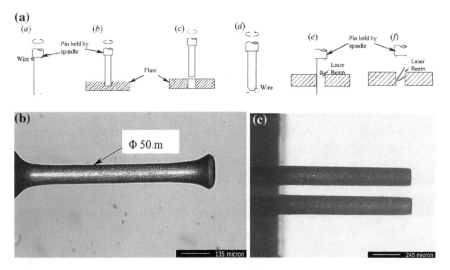

Fig. 4.27 **a** Schematic diagram showing the step-by-step process of micro-EDM–laser compound/sequential process [53]. **b** A single micro-probe fabricated by micro-WEDG process. **c** Pin-plate micro-assembly [micro-joining of two micro-rods that has been fabricated by micro-WEDG] [52]

4.8.2 Micro-EDM-Based Hybrid Processes

4.8.2.1 Micro-EDM and Micro-USM

This hybrid machining process combines micro-EDM and micro-ultrasonic vibration machining (MUSM). The material removal is from both the electrical discharging and mechanical polishing of abrasive slurry. The high-frequency pumping action of the vibrating surface of the electrode accelerates the slurry circulation, giving smaller machining times. The pressure variations in the gap lead to more efficient discharges, which remove more melted metal. The affected layer is reduced, thermal residual stresses are modified, less micro-cracks are observed, and fatigue resistance is increased, due to abrasive action of slurries. The MRR and surface finish of the process depend on the size of the abrasive particles used in MUSM. [55]. Figure 4.28a presents the working principle of micro-EDM and USM hybrid process. The machining of micro-hole with improved inner surface produced by micro-EDM and micro-EDM–USM hybrid processes is presented in Fig. 4.28b–e [55].

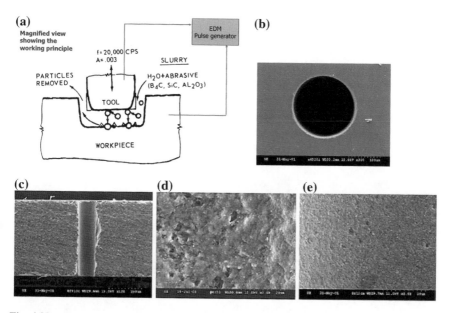

Fig. 4.28 **a** Working principle of micro-EDM and USM hybrid process. **b** *Top* view of the fabricated micro-hole by the hybrid process. **c** Cross-section of the hole shown in fig. **b**, **d** the inner surface of the hole machined by micro-EDM only. **d** The inner surface of the hole machined by micro-EDM and USM hybrid process [55]

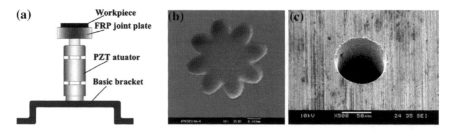

Fig. 4.29 a Working principle of workpiece-vibration-assisted micro-EDM. b Fabrication of blind micro-structure/micro-mold using workpiece-vibration-assisted micro-EDM [57]. c Machining of high-aspect-ratio micro-holes [aspect ratio 16.7] in difficult-to-cut tungsten carbide using workpiece-vibration-assisted micro-EDM [58]

4.8.2.2 Vibration-Assisted Micro-EDM

The process is a combination of micro-EDM and vibration to the workpiece or electrode at the same time. The process improves the flushing conditions and the removal of debris and improves the machining stability and thus reduces the machining time significantly. The process is suitable for deep-hole drilling in hard and difficult-to-cut materials. Depending on the experimental design and objective, the vibration can be applied to the tool electrode [56] or workpiece [57, 58]. The vibration can be low-frequency vibration or ultrasonic vibration. Tool vibration is comparatively more difficult to apply in micro-EDM, as the tool electrode is only of several micron diameters; hence, there is a chance for tool deflection. Therefore, in recent years, research has been carried out on the feasibility of workpiece-vibration-assisted EDM for the fabrication of micro-parts, especially during deep-hole drilling [57, 58]. Figure 4.29 shows the working principle of workpiece vibration and two different micro-structures fabricated by workpiece-vibration-assisted micro-EDM [57, 58].

4.8.2.3 Powder-Mixed Micro-EDM

In recent years, to improve the quality of the micro-EDMed surface and also to reduce the surface defects, several investigators have found this process effective [59]. In this hybrid process, the electrically conductive or semi-conductive powder is mixed in the dielectric, which reduces the insulating strength of the dielectric fluid and increases the spark gap between the tool and workpiece. Enlarged spark gap makes the flushing of debris easier. As a result, the process becomes stable, improving the MRR and surface finish [60]. The sparking is uniformly distributed among the powder particles in the spark gap, thus reducing the intensity of a single spark, which results in uniform shallow craters instead of a single broader crater. Thus, the surface finish improves. There may be some abrasive actions of the powder particles also during the finishing, which reduce the crater boundaries, thus making the surface shiny. Figure 4.30a shows the working principle of

Fig. 4.30 **a** Schematic showing the working principle of powder-mixed micro-EDM process [59]. **b** Machined surface without powder-mixed dielectric. **c** Surface obtained in micro-EDM with powder-mixed dielectric [60]

powder-mixed micro-EDM [59]. The improvement in surface finish and reduction of crater sizes in powder-mixed micro-EDM are presented in Fig. 4.30b–c [60].

4.8.2.4 Micro-Electro-chemical Discharge Machining

The micro-electro-chemical discharge machining (micro-ECDM) process involves a complex combination of the electro-chemical (EC) reaction and electro-discharge (ED) action. The electro-chemical action helps in the generation of the positively charged ionic gas bubbles, for example hydrogen. The electrical discharge action takes place between the tool and the workpiece due to the breakdown of the insulating layer of the gas bubbles as the DC power supply voltage is applied between the tool (or cathode) and the anode, resulting in material removal due to melting, vaporization of the workpiece material, and mechanical erosion [61]. An expanded version of micro-ECDM with conductive powder-mixed electrolyte has been found to produce improved surface finish and integrity compared to micro-ECDM process alone [62]. The working principle of the ECDM system is presented schematically in Fig. 4.31a [63]. Figure 4.31b–c shows a comparison of the surface finish with ECDM and abrasive mixed ECDM [64].

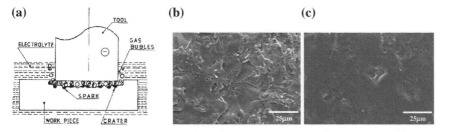

Fig. 4.31 **a** Working principle of the micro-ECDM process [63]. **b** Machining of non-conductive Pyrex glass using micro-ECDM process (Ra: 1.8 μm). **c** Machining of same Pyrex glass using micro-ECDM process with SiC powder (Ra: 1.0 μm) [64]

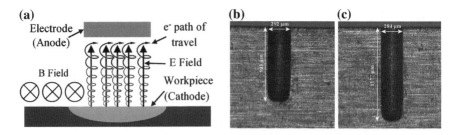

Fig. 4.32 a Working principle of magnetic field–assisted micro-EDM [65]. **b** Cross-section of a micro-hole machined by conventional micro-EDM. **c** Cross-section of micro-hole machined by magnetic field–assisted micro-EDM at the same machining conditions of Fig. **b** [66]

4.8.2.5 Magnetic field–Assisted Micro-EDM

This hybrid process involves a complex combination of micro-EDM and magnetic field assistance in order to improve machining performance by the removal of debris and increasing the MRR. The magnetic field has been introduced to improve debris circulation [65, 66]. Implementing magnetic force perpendicular to the electrode's rotational force produces a resultant force that is efficient in transporting debris out of the hole during machining. A debris particle in a magnetic field–assisted micro-EDM is subjected to two kinds of forces, namely the magnetic force and the centrifugal force. The resultant force on the debris particle is given by the vector addition of the magnetic force and centrifugal force, which helps to flush out the debris particles from the machine zone, thus improving the machining stability, MRR, reducing tool wear, and improving overall micro-EDM performance [65]. Magnetic field–assisted micro-EDM can produce higher-aspect-ratio holes compared with the conventional micro-EDM process under similar working conditions [66]. The application of magnetic fields helps in gap cleaning in micro-EDM due to increased debris transport out of the gap. The enhanced debris removal due to the application of magnetic fields leads to an increase in MRR. It has been reported that for a magnetic material, the MRR increased nearly three times that of a hole cut without the magnetic field [66]. Figure 4.32a shows the schematic representation of magnetic field–assisted micro-EDM [65]. The increment in aspect ratio of micro-hole in magnetic field–assisted micro-EDM can be understood from Fig. 4.32b–c [66].

4.9 Micro-EDM Compared to Other Machining Processes

4.9.1 Compatibility for Materials

Micro-EDM has the advantage of machining of all types of conductive materials such as metals, metallic alloys, graphite, or even some ceramic materials, of

Table 4.1 Compatibility of micro-machining technologies with different materials [68]

Micro-machining technology	Feasible materials
LIGA	Metals, polymers, ceramic materials
Etching	Metals, semiconductors
Excimer LASER	Metals, polymers, ceramic materials
Micro-milling	Metals, polymers
Diamond cutting	Non-ferro metals, polymers
Micro-stereolithography	Polymers
Micro-EDM	Metals, semiconductors, ceramics

whatsoever hardness [67]. Table 4.1 gives an overview of the compatibility of the different micro-machining technologies [68].

4.9.2 Minimum Feature Size, Dimension, and Aspect Ratio

For the evaluation of a micro-machining process, the minimum achievable feature size is very important to consider, as the evolution of micro-machining arises from the need of miniaturization. Table 4.2 presents a comparative evaluation of the micro-machining processes conducted by Makino [69]. Moreover, Fig. 4.33 presents a comparison of the capability with different machining and EDM including accuracy and the surface finish of the final product [70]. The comparison of micro-EDM with other micro-machining processes based on the aspect ratio and dimensions of the micro-features is presented in Fig. 4.34.

4.9.3 Advantages of Micro-EDM Over Other Processes

Compare to traditional micro-machining technologies micro-EDM has several advantages such as [68]

- Micro-EDM requires a low installation cost compared to lithographic techniques.
- Micro-EDM is very flexible, thus making it ideal for prototypes or small batches of products with a high added value.
- Micro-EDM requires little job overhead (such as designing masks).
- Micro-EDM can easily machine complex (even 3D) micro-shapes.
- Shapes that prove difficult for etching are relatively easy for micro-EDM.

From the above discussion, the following are some of the distinct features and applications of micro-EDM compared to other micro-machining processes:

Table 4.2 Evaluation of micro-machining processes by Makino [69]

Process	Principle	Minimum feature size (μm)	Major advantages	Major disadvantages
Micro-molding and casting	Solidification	500	Mass production	Spring back
Micro-punching	Plastic deformation	50	Mass production	Need uniform clearance
Micro-milling	Force	25	Good accuracy and finish	Tool deflection or damage
Micro-grinding	Force	25	Good surface finish	Tool damage
Micro-stereolithography	Lamination	12	Complex 3D shape	Limited work materials
Excimer laser	Ablation	10	No heat damage	Limited work materials
Micro-EDM	Melting/vaporization	5	Negligible force	Surface defect, low MRR
Laser micro-machining	Melting/vaporization	5	No contact/no force	Surface damage
Focused ion beam (FIB)	Sputtering	0.2	Stress free	Low MRR

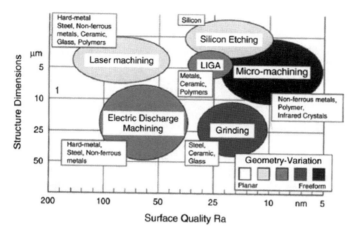

Fig. 4.33 EDM and micro-EDM relative to other machining processes [70]

Fig. 4.34 Micro-EDM relative to other micro-machining processes based on aspect ratio and dimensions of the micro-features

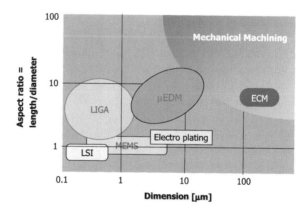

- Micro-EDM has ability to machine any conductive material irrespective of their mechanical hardness. The micro-EDM process can machine materials such as quenched steel and carbides that are mainly used for making cutting tools owing to their very high hardness, and these materials are very difficult to machine using mechanical cutting processes. Micro-EDM can also process materials such as silicon and ferrite, which have high specific resistance.
- The micro-EDM system is designed to maintain a gap between the tool and the workpiece in order to ensure electric discharge between them. Therefore, machining of material can be done without applying pressure on the material, including high-precision machining on curved surfaces, inclined surfaces, and very thin sheet materials which are difficult to drill. Moreover, micro-parts actually used in micro-machines are extremely small, and non-contact machining is particularly very important for them.

- High-aspect-ratio machining can be done using the process. In an ordinary perforating process, micro-EDM can easily perforate a hole to a depth equivalent to five times the bore diameter.
- High-precision and high-quality machining can be done. The shape of the tool electrode and the ED gap between the electrode and the workpiece determine precision of the machined shape. Moreover, the micro-EDM produces very small burrs, much smaller than those seen in mechanical drilling and milling operations, and therefore does not need subsequent deburring operations.

4.10 Conclusions

The capability of machining intricate micro-features with sub-micron-level dimensional accuracy in difficult-to-cut material has made the micro-EDM process an inevitable and one of the most popular non-conventional micro-machining processes. To establish micro-EDM as an effective process and to overcome the shortcomings of the process, continuous research and developments are being carried out. Future advancement in the area of micro-EDM will continue toward understanding the fundamental science and engineering of the process as well as broadening the application of the process for the industries. The automation of the micro-EDM process will be one of the key focuses for future research and development, which holds the micro-EDM back to be used widely extensively in industries at present. The automation in micro-EDM will improve the batch production of micro-parts in industries. Some other important areas for the future development are fabrication techniques of on-machine tool fabrication, improvement of in situ measurement of the dimensions and accuracy of the fabricated parts, development of new hybrid and compound micro-machining technologies, and so on. The development of multipurpose miniature machine tool can significantly improve the advanced research on the areas of micro-EDM and micro-EDM-based compound and hybrid micro-machining processes.

References

1. Ho KH, Newman ST (2003) State of the art electrical discharge machining (EDM). Int J Mach Tools Manuf 43:1287–1300
2. Ho KH, Newman ST, Rahimifard S, Allen RD (2004) State of the art wire electrical discharge machining. Int J Mach Tools Manuf 44:1247–1259
3. Arbizu ID, Pérez CJL (2013) History and foundations of EDM: the least conventional of the machining. http://www.interempresas.net/MetalWorking/Articles/12068-The-least-conventional-of-the-machining.html, last accessed 24 Apr 2013
4. Kurafuji H, Masuzawa T (1968) Micro-EDM of cemented carbide alloys. Jpn Soc Electr Mach Eng 2(3):1–16

5. Jameson EC (2001) Description and development of electrical discharge machining (EDM), in: electrical discharge machining. Society of Manufacturing Engineers, Dearborn, Michigan p 12
6. Masuzawa T (2000) State of the art micromachining. Ann CIRP 49(2):473–488
7. Gentili E, Tabaglio L, Aggogeri F (2005) Review on micromachining techniques. Courses Lectures Int Centre Mech Sci 486:387 dimgruppi.ing.unibs.it
8. Schumacher BM (2004) After 60 years of EDM the discharge process remains still disputed. J Mater Process Technol 149:376–381
9. Takahata K (2009) Micro-electro-discharge machining technologies for MEMS, Chapter 10. Micro Electron Mech Syst INTECH, Croatia, pp 144–164
10. Rahman M, Asad ABMA, Masaki T, Saleh T, Wong YS, Kumar AS (2010) A multiprocess machine tool for compound micromachining. Int J Mach Tools Manuf 50(4):344–356
11. Wong YS, Rahman M, Lim HS, Han H, Ravi N (2003) Investigation of micro-EDM material removal characteristics using single RC-pulse discharges. J Mater Process Technol 140(1–3):303–307
12. Masuzawa T (2001) Micro EDM. In: Proceedings of the ISEM XIII. Fundación Tekniker, pp 3–19 ISBN 932064-0-7
13. Micro manufacturing, differences between macro and micro EDM, Ministerio de Ciencia e Innovación. http://www.micromanufacturing.net/didactico/Desarollo/edm-eng/6-differences-between-macro-and-micro-edm. Last Accessed 10 Dec 2012
14. Katz Z, Tibbles CJ (2005) Analysis of micro-scale EDM process. Int J Adv Manuf Technol 25:923–928
15. Uhlmann E, Piltz S, Doll U (2005) Machining of micro/miniature dies and moulds by electrical discharge machining—recent development. J Mater Process Technol 167:488–493
16. Han F, Yamada Y, Kawakami T, Kunieda M (2004) Improvement of machining characteristics of micro-EDM using transistor type isopulse generator and servo feed control. Precis Eng 28:378–385
17. Kunieda M, Lauwers B, Rajurkar KP, Schumacher BM (2005) Advancing EDM through fundamental insight into the process. Ann CIRP 54(2):599–622
18. Rajurkar KP, Levy G, Malshe A, Sundaram MM, McGeough J, Hu X, Resnick R, De Silva A (2006) Micro and nano machining by electro-physical and chemical processes. Ann CIRP 55(2):643–666
19. Zhang L, Jia Z, Liu W, Li A (2012) A two-stage servo feed controller of micro-EDM based on interval type-2 fuzzy logic. Int J Adv Manuf Technol 59(5–8):633–645
20. Yang GH, Liu F, Lin HB (2011) Research on an embedded servo control system of micro-EDM. Appl Mech Mater 120:573–577
21. Rajurkar KP, Wang WM (1991) On-line monitor and control for wire breakage in WEDM. Ann CIRP 40(1):219–222
22. Snoeys R, Dauw D, Kruth JP (1980) Improved adaptive control system for EDM processes. Ann CIRP 29(1):97–101
23. McGeough JA (1988) Advanced methods of machining, 1st edn. Chapman and Hall, USA. ISBN 0-412-31970-5
24. Pham DT, Dimov SS, Bigot S, Ivanov A, Popov K (2004) Micro-EDM—recent developments and research issues. J Mater Process Technol 149:50–57
25. Puertas I, Luis CJ, Álvarez L (2004) Analysis of the influence of EDM parameters on surface quality, MRR and EW of WC–Co. J Mater Process Technol 153, 154:1026–1032
26. Hewidy MS, El-Taweel TA, El-Safty MF (2005) Modelling the machining parameters of wire electrical discharge machining of Inconel 601 using RSM. J Mater Process Technol 169:328–336
27. Qu J, Shih AJ (2002) Development of the cylindrical wire electrical discharge machining process, part 2: surface integrity and roundness. Trans ASME 124
28. Tsai Y, Masuzawa T (2004) An index to evaluate the wear resistance of the electrode in micro-EDM. J Mater Process Technol 149:304–309
29. Cao G, Zhao W, Wang Z, Guo Y (2005) Instantaneous fabrication of tungsten microelectrode based on single electrical discharge. J Mater Process Technol 168:83–88

30. Jahan MP, Wong YS, Rahman M (2009) A study on the quality micro-hole machining of tungsten carbide by micro-EDM process using transistor and RC-type pulse generator. J Mater Process Technol 209(4):1706–1716

31. Bleys P, Kruth JP, Lauwers B (2004) Sensing and compensation of tool wear in milling EDM. J Mater Process Technol 149:139–146

32. Yu Z, Jun T, Kunieda M (2004) Dry EDM of cemented carbide. J Mater Process Technol 149:353–357

33. Tao J, Shih AJ, Ni J (2008) Near-dry EDM milling of mirror-like surface finish. Int J Electr Mach 13:29–33

34. Subbu SK, Karthikeyan G, Ramkumar J, Dhamodaran S (2011) Plasma characterization of dry μ-EDM. Int J Adv Manuf Technol 56(1–4):187–195

35. Zhang QH, Zhang JH, Deng JX, Qin Y, Niu ZW (2002) Ultrasonic vibration electrical discharge machining in gas. J Mater Process Technol 129:135–138

36. Lim HS, Wong YS, Rahman M, Lee EMK (2003) A study on the machining of high-aspect ratio micro-structures using micro EDM. J Mater Process Technol 140:318–325

37. Jahan MP, Rahman M, Wong YS, Fuhua L (2010) On-machine fabrication of high-aspect-ratio micro-electrodes and application in vibration-assisted micro-electrodischarge drilling of tungsten carbide. In: Proceedings of the institution of mechanical engineers, part B: J Eng Manuf 224(5):795–814

38. Masuzawaa T, Fujinoa M, Kobayashia K, Suzukib T, Kinoshita N (1985) Wire electro-discharge grinding for micro-machining. CIRP Ann Manuf Technol 34(1):431–434

39. Song KY, Chung DK, Park MS, Chu CN (2009) Micro-electrical discharge drilling of tungsten carbide using deionized water. J Micromech Microeng 19:045006

40. Egashira K, Taniguchi T, Hanajima S (2006) Planetary EDM of micro holes. Int J Electr Mach 11:15–18

41. Yu ZY, Rajurkar KP, Shen H (2002) High aspect ratio and complex shaped blind micro holes by micro EDM. Ann CIRP 51(1):359–362

42. Kim BH, Park BJ, Chu CN (2006) Fabrication of multiple electrodes by reverse EDM and their application in micro ECM. J Micromech Microeng 16(4):843–850

43. Sarix Micro-EDM machine. http://www.sarix.com/index_e.html. Last Accessed 10 Dec 2012

44. Asad ABMA, Masaki T, Rahman M, Lim HS, Wong YS (2007) Tool-based micro-machining. J Mater Process Technol 192–193:204–211

45. Morgan CJ, Vallance RR, Marsh ER (2006) Micro-machining and micro-grinding with tools fabricated by micro electro-discharge machining. Int J Nanomanuf 1(2):242–258

46. Wada T, Masaki T, Davis DW (2002) Development of micro grinding process using micro EDM trued diamond tools. In: Proceedings of the annual meeting of ASPE

47. Fleischer J, Masuzawa T, Schmidt J, Knoll M (2004) New applications for micro-EDM. J Mater Process Technol 149:246–249

48. Campana S, Miyazawa S (1999) Micro-EDM and ECM in DI water. In: Proceedings of annual meeting of American society of precision engineering (ASPE)

49. Nguyen MD, Rahman M, Wong YS (2012) Simultaneous micro-EDM and micro-ECM in low-resistivity deionized water. Int J Mach Tools Manuf 54–55:55–65

50. Zeng Z, Wang Y, Wang Z, Shan D, He X (2012) A study of micro-EDM and micro-ECM combined milling for 3D metallic micro-structures. Precis Eng 36(3):500–509

51. Takahata K, Shibaike N, Guckel H (2000) High-aspect-ratio WC-Co microstructure produced by the combination of LIGA and micro-EDM. Microsyst Technol 6(5):175–178

52. Huang J-D, Kuo C-L (2002) Pin-plate micro assembly by integrating micro-EDM and Nd-YAG laser. Int J Mach Tools Manuf 42:1455–1464

53. Kuo C-L, Huang J-D, Liang H-Y (2002) Precise micro-assembly through an integration of micro-EDM and Nd-YAG. Int J Adv Manuf Technol 20:454–458

54. Kuo C-L, Huang J-D, Liang H-Y (2003) Fabrication of 3D metal microstructures using a hybrid process of micro-EDM and laser assembly. Int J Adv Manuf Technol 21:796–800

55. Lin YC, Yan BH, Chang YS (2000) Machining characteristics of titanium alloy (Ti-6Al-4V) using a combination process of EDM with USM. J Mater Process Technol 104(3):171–177

56. Endo T, Tsujimoto T, Mitsui K (2008) Study of vibration-assisted micro-EDM-the effect of vibration on machining time and stability of discharge. Precis Eng 32(4):269–277
57. Tong H, Li Y, Wang Y (2008) Experimental research on vibration assisted EDM of microstructures with non-circular cross-section. J Mater Process Technol 208(1–3):289–298
58. Jahan MP, Saleh T, Rahman M, Wong YS (2010) Development, modeling, and experimental investigation of low frequency workpiece vibration-assisted micro-EDM of tungsten carbide. J Manuf Sci Eng 132(5):054503 (8)
59. Kansal HK, Singh S, Kumar P (2007) Technology and research developments in powder mixed electric discharge machining (PMEDM). J Mater Process Technol 184:32–41
60. Jahan MP, Anwar MM, Wong YS, Rahman M (2009) Nanofinishing of hard materials using micro-EDM. Proc Inst Mech Eng Part B J Eng Manuf 223:1127–1142
61. Sorkhel SK, Bhattacharyya B, Mitra S, Doloi B (1996) Development of electrochemical discharge machining technology for machining of advanced ceramics. In: International conference on agile manufacturing, pp 98–103
62. Han M-S, Min B-K, Lee SJ (1999) Improvement of surface integrity of electro-chemical discharge machining process using powder-mixed electrolyte. J Mater Process Technol 95:145–154
63. Bhattacharyya B, Doloi BN, Sorkhel SK (1999) Experimental investigations into electrochemical discharge machining (ECDM) of non-conductive ceramic materials. J Mater Process Technol 95:145–154
64. Yang CT, Song SL, Yan BH, Huang FY (2006) Improving machining performance of wire electrochemical discharge machining by adding SiC abrasive to electrolyte. Int J Mach Tools Manuf 46:2044–2050
65. Heinz K, Kapoor SG, Devor RE, Surla V (2011) An investigation of magnetic-field-assisted material removal in micro-EDM for nonmagnetic materials. J Manuf Sci Eng 133(021002):9
66. Yeo SH, Murali M, Cheah HT (2004) Magnetic field assisted micro electro-discharge machining. J Micromech Microeng 14:1526–1529
67. Jahan MP, Rahman M, Wong YS (2011) A review on the conventional and micro-electrodischarge machining of tungsten carbide. Int J Mach Tools Manuf 51(12):837–858
68. Reynaerts D, Heeren PH, Van Brussel H, Beuret C, Larsson O, Bertholds A (1997) Microstructuring of silicon by electro-discharge machining (EDM) part II: applications. Sens Actuators 61:379–386
69. Grzesik W (2008) Advanced machining processes of metallic materials: theory, modelling and applications chapter sixteen micromachining. Elsevier publishers, Amsterdam
70. Byrne G, Dornfeld D, Denkena B (2003) Advancing cutting technology. Ann CIRP 52(2):483–507

Chapter 5
Prototype Machine for Micro-EDM

Ivo M. F. Bragança, Gabriel R. Ribeiro, Pedro A. R. Rosa
and Paulo A. F. Martins

Abstract This chapter presents constructive details for a micro-electrical discharge machine that is adequate to study the fundamentals of the process and to educate upcoming engineers in the latest industrial technologies. The machine was designed, fabricated, and instrumented by the authors and consists of a rigid structure, electrical and electronic components, a dielectric flowing system, and positioning control devices. The machine operates with low voltage, low energy, and high-frequency short electrical pulses and makes use of tool electrodes that are capable of drilling holes with diameters in the micrometer range. The presentation provides constructive details for those readers who may be interested in developing an in-house μEDM and puts emphasis on its adequacy for investigating the influence of operative parameters on electrical spark discharges, morphology of craters, and material removal mechanisms.

5.1 Introduction

Electrical discharge machining (EDM) is a material removal process based on controlled erosion of metals by the intense heat of electric spark discharges. There are three main variants of EDM; conventional (or sinker) EDM, wire EDM, and hole EDM drilling.

In conventional EDM, the workpiece is immersed into a dielectric (electrically non-conducting) fluid and connected to a terminal of a DC power supply while the

Ivo M. F. Bragança · G. R. Ribeiro · P. A. R. Rosa · P. A. F. Martins (✉)
IDMEC, Instituto Superior Técnico, Technical University of Lisbon,
Av. Rovisco Pais 1049-001 Lisbon, Portugal
e-mail: pmartins@ist.utl.pt

Ivo M. F. Bragança
e-mail: ivo.braganca@ist.utl.pt

J. P. Davim (ed.), *Nontraditional Machining Processes*,
DOI: 10.1007/978-1-4471-5179-1_5, © Springer-Verlag London 2013

tool-electrode (hereafter called 'electrode') is immersed at some distance (hereafter called 'gap') from the workpiece and connected to the other terminal (see Fig. 5.1a). The first machines were developed in the mid-1950s and the main applications of conventional EDM are the fabrication of blind cavities and hole drilling using electrodes machined from graphite or copper to the desired shape of the cavities and holes.

The development of CNC systems in the 1970s brought dramatic improvements to EDM technology that lead to significant gains in accuracy, quality, productivity, and earnings. One of the key innovations of that period was the development of wire EDM that makes use of computer numerical control (CNC) systems to cut extremely complicated shapes, automatically, precisely, and economically (Fig. 5.1b).

The working principle of wire EDM is similar to that of conventional EDM, but instead of using and electrode that slowly plunges into the workpiece, it uses a traveling wire electrode (made from copper, brass, or molybdenum with diameters ranging from 0.01 to 0.5 mm) that passes through the workpiece to remove material. In order to feed the wire electrode through the workpiece, an initial hole must often be drilled in the workpiece prior to wire EDM.

The third variant of EDM is used for drilling of holes in any electrical conductive material, whether hard or soft, including tungsten carbide (Fig. 5.1c). The term 'fast hole EDM drilling' is used to distinguish the process from conventional EDM which can also be used for drilling holes but in much slower speeds. The working principle of fast hole EDM drilling is similar to that of conventional EDM, and major differences are related to the fact that electrodes rotate and are hollow. The rotating electrode helps ensuring better concentricity and reducing wear, while the hollow features allow dielectric fluid to flow through the electrode directly to the working gap.

EDM technology offers significant advantages against conventional machining due to its capability of fabricating complex blind cavities, deep holes, and extremely complicated cutting shapes without mechanical contact between the electrode and the workpiece and regardless of material hardness [1].

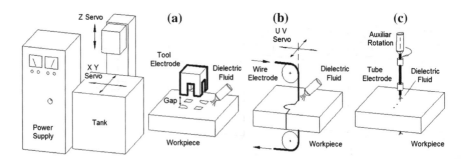

Fig. 5.1 Three main variants of electrical discharge machining (EDM) **a** conventional or sinker EDM, **b** wire EDM, and **c** fast hole EDM drilling

The above-mentioned advantages, together with the flexibility and short manufacturing times of EDM in small batch production, have been focusing attention of researchers and industrial companies engaged with the production of micro-mechanical parts to the possibility of taking EDM technology a step forward to meet the precision requirements of tool making in the micrometer range [2]. The growing demand on micro-mechanical parts is driven by a global trend toward the miniaturization of products originated not only from consumers, who want small and highly sophisticated electronic equipment, but also from recent applications in medicine, electronics, and optoelectronics.

However, EDM technology and its machine-tools cannot be simply scaled down to be applied on the micro-level. Scaling down requires combining information on key-operating parameters of EDM at micro-level and their influence in the morphology of the craters with new solutions and concepts for the design and fabrication of micro-electrical discharge machine.

In the past years, several well-known manufacturers started to offer sophisticated, high-performance, μEDMs capable of producing very fine finishes and fabricating very small features in micro-tools such as deep holes with up to 30 μm of diameter. μEDM is one of the most promising technologies in terms of size and precision [3]. However, commercial μEDMs are generally expensive and very often do not provide key functionalities that are needed for education and research purposes. Commercially available μEDMs commonly prevent free selection of the operative parameters (such as duty cycle and frequency of electrical pulses) and are not designed for generating single and different types of electrical spark discharges as it is necessary for investigating the fundamentals of μEDM, namely the material removal mechanisms. A state-of-the-art in the field is provided by Rajurkar et al. [4].

From what was mentioned before, there is great interest in developing low-cost μEDM laboratory machines that can be used for educational and research objectives. This chapter is concerned with the above-mentioned gap between developers and users of μEDM and is organized in two main parts. The first part aims to provide constructive details of a μEDM prototype machine that is capable of generating single or multiple electrical spark discharges. The machine was designed, fabricated, and instrumented by the authors and is capable of drilling holes with dimensions in the micrometer range. The second part of the chapter analyzes how operative parameters such as the electrode diameter, frequency, and voltage influence electrical spark discharges and material removal mechanisms, namely material removal rate (MRR) and morphology of the craters.

5.2 Review of the Fundamentals of EDM

When sufficient voltage is applied between the electrode and the workpiece, the dielectric fluid ionizes and forms a plasma channel that melts and vaporizes the material located on the surface of the workpiece. Once the electrical spark discharge is complete, the vaporized cloud solidifies and the resulting tiny solid material particles (debris) are removed from the gap between the electrode and workpiece by flowing dielectric fluid (Fig. 5.2).

The main operative parameters of EDM are as follows: (1) the polarity between the electrode and the workpiece, (2) the open-circuit voltage, (3) the frequency and intensity of the electric current pulses, (4) the tool or wire electrodes, and (5) the dielectric fluids.

In what concerns polarity, the common practice in EDM is to make electrode negative and workpiece positive (direct polarity) in order to achieve higher metal removal rate. However, researches showed that reverse polarity, in which electrode is positive and workpiece is negative, may be a better option for diminishing roughness and improving surface quality of the workpiece [5].

The open-circuit voltage is the critical ionization voltage below which there is no electrical spark discharge and, therefore, no electricity flowing between the electrode and the workpiece. The discharge energy during EDM is provided by a direct current pulse power generator system (RC or transistorized), and the frequency is characterized by the pulse on-time t_{on}, the pulse off-time t_{off}, the period $T = t_{on} + t_{off}$ (seconds), and the duty cycle D, which is defined as follows:

$$D = \frac{t_{on}}{t_{on} + t_{off}} \tag{5.1}$$

The average discharge current $\bar{I}$ (Amperes) in the period T is calculated from the amplitude of the pulse current I (also known as the 'peak discharge current') as follows:

$$\bar{I} = \frac{t_{on}}{t_{on} + t_{off}} \cdot I \tag{5.2}$$

Fig. 5.2 Schematic representation of the plasma channel, vaporized material cloud, solid material particles removed from the workpiece surface and wash out by flowing dielectric fluid. The symbols t_{on} and t_{off} denote the pulse on-times and pulse off-times, respectively

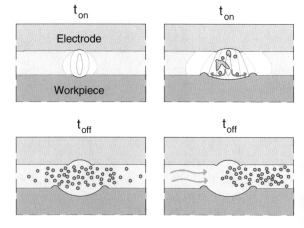

The MRR (mm^3/h) in EDM is primarily a function of the average discharge current $\bar{I}$ (Eq. 5.2) and melting temperature T_m (Celsius) of the workpiece material,

$$\text{MRR} = C \cdot \bar{I} \cdot T_m^a \tag{5.3}$$

where C and a are process and material constants that depend on the material and geometry of the electrodes, on the type and flow rate of dielectric fluids and on the nature of electrical spark discharges, among other parameters.

Electrical spark discharges are commonly classified into four different categories: (1) gap open, (2) normal sparks, (3) arc sparks and (4) short-circuits, and research work is currently ongoing to determine the relation between the type of sparks, MRR, and surface finishes [6–8]. The possibility of generating single electrical spark discharges in a µEDM laboratory machine is crucial to analyze the different types of sparks, to provide a new level of understanding on plasma formation in the dielectric and to explain how these phenomena influence material removal mechanisms. In fact, as concluded by Schumacher [9] and Kunieda et al. [10], there are fundamental concepts of EDM that needs to be further investigated because they are not properly well understood. Some of these concepts will be addressed later in the presentation.

5.3 The µEDM Prototype Machine

Figure 5.3 shows a schematic representation of the µEDM experimental apparatus that was developed by the authors for educational and research purposes. The design of the µEDM prototype machine follows some of the issues, trends, and needs for the development of equipment for electrophysical micro-machining that were previously identified by Rajurkar et al. [4].

For purposes of presentation, and although not corresponding to what readers directly observe in Fig. 5.3, the µEDM prototype machine will be split into four broad groups of components: (1) basic structure, (2) electrical circuit, (3) dielectric flowing system, and (4) position control devices.

5.3.1 Basic Structure

The primary function of the structure is to support and position the tool-electrode and workpiece. The µEDM prototype machine was built upon a robust C-frame structure which consists of a base, an X–Y table, a column, and a servo head (Fig. 5.4).

The X–Y table includes means of attaching the workpiece to the table surface and allows moving the workpiece into position. The servo head holds the electrode

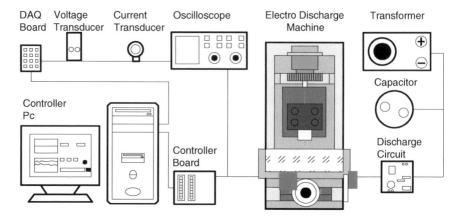

Fig. 5.3 Schematic representation of the µEDM experimental apparatus

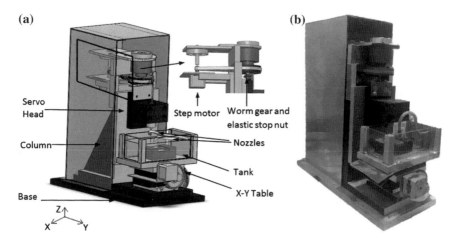

Fig. 5.4 a Schematic drawing and **b** photograph of the uEDM prototype

and is reinforced with lateral guides to reduce vibrations and undesirable side motion. Lateral guides ensure the electrode to move accurately over the entire length of travel and to be precisely feed into the workpiece.

The C-frame structure of the µEDM combines rigidity with flexibility and slim design. Rigidity is necessary for maintaining very precise control over the working gap between electrode and workpiece. Flexibility allows easy changing of components and integration of new accessories. The slim design concept offers excellent accessibility to the working area from the front and the sides.

5.3.2 Electrical Circuit

The electrical circuit of the proposed μEDM is different from those currently utilized in commercial EDM because it allows generating single and multiple electrical spark discharges. The modular construction of the electrical circuit comprises a power system with a main bank of capacitors and easily interchangeable discharge circuits that quickly adapt the machine to meet specific requirements of education and research (see Fig. 5.5).

The EDM power system consists of a variable-voltage transformer, a constant current rectifier for converting the single-phase AC supplied with 220 V to DC with voltage in the range 20–330 V and current rating up to 4 A and a main bank of capacitors (Fig. 5.5a).

Figure 5.5b–d shows three different interchangeable discharge circuits. The circuit shown in Fig. 5.5b (named 'impulse generator') makes use of a variable resistance and a metal–oxide semiconductor field effect transistor (abbreviately named as 'MOSFET') to control discharge and to generate square pulses of electric current with frequencies in the range 1 Hz–500 kHz over a duty cycle $D = 0.5$. The impulse generator is utilized both in single- and multiple-electrical-spark-discharge modes. When the μEDM is set to single-spark mode, the MOSFET behaves as a switch that opens and closes the discharge circuit by supplying control voltage and pulse on-time to a PCI-MIO-16E board from National

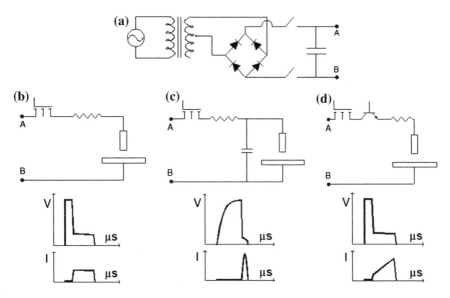

Fig. 5.5 Scheme of the electrical circuit of the μEDM prototype machine **a** power system consisting of a variable-voltage transformer, a constant current rectifier and a main bank of capacitors, **b** discharging circuit consisting of a variable resistance and a transistor, **c** discharging circuit consisting of a RC circuit and a transistor, and **d** discharging circuit consisting of a variable resistance, a pulse generator, and a transistor

Instruments. The variable resistance in the impulse generator circuit controls the current and, therefore, the total discharge energy available for the sparks.

Changing the impulse generator by the discharge circuit shown in Fig. 5.5c allows the user to utilize a 'RC-type relaxation generator' to create and control square pulses of electric current. In the RC relaxation generator, the discharge energy comes from a capacitor that is connected in parallel with the machining gap. As a result of the low impedance of the plasma channel, the ionization time is larger (although the peak voltage is the same), the discharge duration is shorter, and the discharge current is higher than in the 'impulse generator' discharge circuit (refer to the graphics included in Fig. 5.5b, c).

The circuit shown in Fig. 5.5d is named as 'electronic pulse generator' and is a modification of the impulse generator (Fig. 5.5b) to allow studying the influence of current flow with time in single electrical spark discharges.

All the electrical circuits shown in Fig. 5.5b–d allow users to control the discharge frequency by adjusting the on-time and off-time of the transistors that control the pulses of electric current.

5.3.3 Dielectric Flowing System

The dielectric flowing system consists of a dielectric fluid, a pump, a filter; two nozzles, and a tank (refer to Fig. 5.4). The dielectric fluid submerges the workpiece and acts as an insulator until the potential difference between the electrode and the workpiece is sufficiently high to produce an electrical spark discharge. The discharge forms a plasma channel in the fluid that removes material particles from the surface of the workpiece by melting and vaporization.

The pump and the maneuverable nozzles, placed on opposite sides of the electrode to minimize pressure-induced oscillations, force the dielectric to circulate through the gap between the electrode and the workpiece, directly to the region being machined, in order to wash out the solid metal particles (debris) and to provide a cooling medium (Fig. 5.2).

The filter helps the dielectric to remain clean by retaining debris and contaminants that may cause short-circuits and ensures that electrical properties of the dielectric (e.g., insulation properties) in the gap remain similar to the original values.

The tank is made from glass reinforced with aluminum due to its lack of chemical affinity with the dielectric fluids commonly utilized in the μEDM prototype machine.

5.3.4 Position Control Devices

The X–Y table is responsible for positioning the workpiece so that μEDM takes place at the proper location. The Z-axis servo head is the machine component that moves the electrode in vertical direction to maintain an adequate electrode-to-workpiece gap distance so that spark discharges will take place. The movement of the X, Y, and Z-axis is obtained from fine pitch worm gears driven by electric stepper motors.

Figure 5.6 shows the console panel of the Lab View computer-based software developed by the authors to monitor and control movement in the X, Y, and Z-axis. The console panel is structured into two different modules. The leftmost module allows manual positioning (by means of pushbuttons) of the X–Y table by specifying the displacement and velocity of each individual axis. The display window allows visualizing the position of the electrode in the X, Y coordinate system and helps predefining and storing additional drilling positions.

The rightmost module of the console panel allows monitoring and controlling the vertical position of the electrode. The moving mechanism of the vertical (Z-axis) is made up of an electric stepper motor, driver and loose pulleys (with a transmission ratio of 1:9), a fine pitch worm gear, an elastic stop nut, a timing belt, and electric sensors (Fig. 5.4).

The elastic stop nut resists loosening under vibrations and torque and the timing belt prevents slippage during transmission of motion (refer to the schematic detail in Fig. 5.4). The Z-axis positioning mechanism is capable of moving the electrode along its length with an accuracy of 1 μm in 100-mm displacement, under proper alignment and tension applied to the timing belt. The electric sensors consist of a Hameg HZ 100-voltage transducer and a Bergoz CTB1.0 current transducer (refer to Fig. 5.3) that monitor the voltage and current between the electrode and workpiece during the electrical spark discharges.

The Z-axis servo head utilizes feedback signals of voltage and current obtained from the previously mentioned sensors to keep the gap constant and prevent the electrode from shorting out against the workpiece. The two display windows included in the rightmost module of the console panel shown in Fig. 5.6 allow

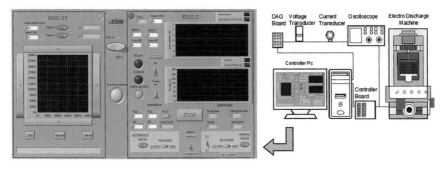

Fig. 5.6 Console panel of the XY table and of the servo-controlled Z-axis

Table 5.1 Main
characteristics of the μEDM
prototype machine

Overall dimensions	230 × 450 × 450 (mm)
Worktable size (X × Y)	25 × 25 (mm)
Worktable travel (X × Y × Z)	25 × 25 × 25 mm
Precision (X-, Y-axis)	0.750 μm
Precision (Z-axis)	0.250 μm
Maximum velocity (X-, Y-axis)	0.35 mm/s
Maximum velocity (Z-axis)	0.05 mm/s
Voltage	20–330 V
Maximum output current	4 A
Maximum frequency	500 kHz
Dielectric flowing rate	0.75 l/min
Maximum electrode diameter	2 mm
Power supply	Single-phase AC, 220 V

monitoring the evolution of voltage and current with time while the available pushbuttons allow changing the gap between the electrode and workpiece in real time.

Table 5.1 resumes the major characteristics of the μEDM prototype machine that was designed and fabricated by the authors.

5.3.5 Data Acquisition

The electric stepper motors utilized in the worm gear position mechanisms produce vibrations that are transmitted to the structure of the μEDM. Knowledge of the range of frequencies associated with these vibrations allows users to select operative parameters where the role played by induced vibrations is minimum. In case of the proposed μEDM prototype machine, the identification and quantification of the aforementioned range of frequencies, where the amplitude of vibrations is significant, were performed by means of a Brüel and Kjær's PULSE platform.

The signals of the Hameg HZ 100 voltage transducer and of the Bergoz CTB 1.0 current transducer utilized in the servo control of the Z-axis are registered in a data acquisition board (National Instruments PCI-6115 with BNC-2120) and can be visualized on a oscilloscope (Tektronix 2004B) for facilitating real-time analysis.

A computer program developed by the authors is used for the treatment of experimental data, namely for performing the recognition and quantification of the different types of electrical spark discharges that are generated during the μEDM process. This is considered one of the most effective procedures for monitoring electrical discharging machining [11].

Figure 5.7a shows the experimental evolution of voltage and current versus pulse on-time for a real (experimentally acquired) electrical spark discharge.

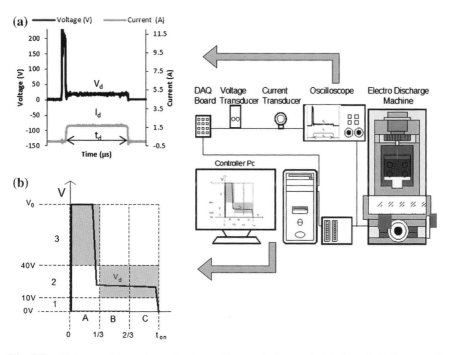

Fig. 5.7 a Voltage and current versus pulse on-time t_{on} during an electrical spark discharge, and **b** automatic recognition of the different types of discharge patterns through the analysis of the experimental evolution of voltage versus pulse on-time t_{on}

The computer program works directly with the experimental evolution of voltage versus pulse on-time $V(t_{on})$ and automatically identifies the regions that lie above and below the curve shown in Fig. 5.7b. Region 3A, for example, corresponds to the ionization potential (open-circuit voltage V_0) while regions 2B and 2C correspond to the discharge potential (working voltage V_d), which varies from approximately 10–40 V in the proposed μEDM prototype machine.

The capability of recognizing different types of discharge patterns in the (A, B, C) × (1, 2, 3) display window allows the computer program to automatically classify sparks as gap open, normal, arc, and short-circuits, to quantify its percentages of occurence and to determine the values of the pulse on-time t_{on}, open-circuit voltage V_0 and working voltage V_d.

5.4 Results and Discussion

This section of the chapter starts by presenting the experimental work plan that was utilized for assessing the performance of μEDM prototype machine for drilling holes with dimensions in the micrometer range and proceeds by analyzing how operative parameters such as the electrode diameter, frequency, and voltage influence the type of electrical spark discharges and material removal mechanisms, namely MRR and morphology of the craters.

5.4.1 Experimental Work Plan

The performance of the μEDM prototype machine was assessed by drilling micro-holes in stainless steel AISI 304 sheets with 1 mm thickness. Electrolytic copper (DIN E-Cu58) wire was utilized as electrode, and Shell Macron EDM 110 oil was used as dielectric.

The experimental work plan was designed in order to isolate the influence of four main process parameters that were considered critical for analyzing the MRR and the morphology of the craters produced with different types of electrical spark discharges (Table 5.2): (1) the diameter of the electrode, (2) the frequency of the square pulses of electric current, (3) the open-circuit voltage, and (4) the operating mode (single- or multiple-spark mode).

The experimental values of the MRR (mm^3/h) were obtained from the ratio between the volume removed from the workpiece and the total drilling time.

By keeping the rest of the parameters at constant value, namely (1) the flow rate of dielectric and (2) the polarity (reverse polarity) between the electrode and the workpiece, it was possible to reduce the total number of variables that influence the process. Otherwise, the number of possible combinations of variables would become quite large.

5.4.2 Vibrations of the Servo Head

The electric stepper motors utilized in the worm gear position mechanisms produce vibrations that are transmitted to the structure of the μEDM. Figure 5.8

Table 5.2 Experimental work plan performed in stainless steel AISI 304 sheets with 1 mm thickness

Electrode Diameter D (mm)	Frequency f (kHz)	Open Circuit Voltage V_0 (V)	Discharge Circuit
0.3 and 1	10, 140 and 200	80, 140 and 200	Impulse generator RC relaxation generator Electronic pulse generator

shows the transmissibility of vibrations that are produced under different working frequencies of the motors to the servo head that holds and moves the electrode. As seen, maximum-induced vibrations on the electrode holder have amplitudes below 150 nm and the range of frequencies where transmissibility is higher is around 100 Hz.

5.4.3 Material Removal Rate

Figure 5.9 shows MRR as a function of the frequency of the pulses and of the open-circuit voltage, for two different electrode diameters. Results were obtained with the impulse generator discharge circuit and show that MRR decreases as the

Fig. 5.8 Experimental displacements of the servo head due to transmissibility of vibrations that are produced under different working frequencies of the electric stepper motors

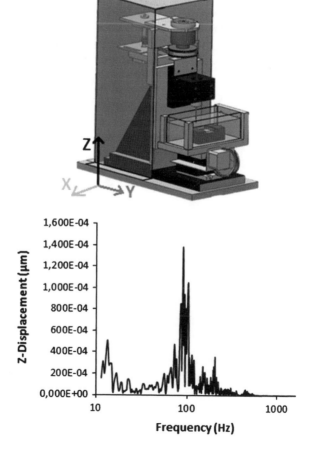

diameter of the electrode decreases. However, the variation of results with frequency and open-circuit voltage requires further analysis.

In case of electrodes with 1 mm diameter, MRR is graphically illustrated as a concave function which increases and then decreases with higher levels of frequency, over a constant duty cycle $D = 0.5$ (Fig. 5.9a). The leftmost region of the graphic (where MRR increases with frequency) is attributed to higher values of current density inside the plasma that is formed in the dielectric. This is because the smallest frequencies give rise to durations of the pulse on-time t_{on} large enough to cause a significant drop in the efficiency of the plasma due to greater losses of thermal energy to the dielectric fluid and to the workpiece.

The rightmost region of the graphic in Fig. 5.9a (where MRR decreases at higher frequencies) is attributed to smaller gaps between the electrode and the

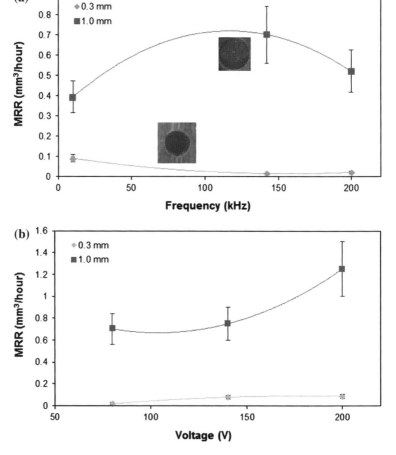

Fig. 5.9 **a** Influence of the frequency of electric current pulses and of, **b** open-circuit voltage in material removal rate (amplitude of current 1.5 A and diameters of the electrodes (0.3 and 1 mm)

workpiece and to smaller durations of the pulse off-time t_{off}. Smaller gaps create difficulties in washing out the debris by the flowing dielectric fluid whereas small durations of the pulse off-time t_{off} cause difficulties in ensuring that properties of the dielectric fluid in the gap recover the original values. The risk of debris being trapped in the gap between the electrode and workpiece does not ensure the electrical insulation properties that are required for a dielectric fluid and gives rise to short-circuiting that significantly decreases the overall surface quality of the micro-holes.

The MRRs of the electrodes with 0.3 mm diameter are practically insensitive to frequency and up to 10 times smaller than those obtained with the electrodes of 1.0 mm diameter (Fig. 5.9a). The overall decrease in MRR is explained by the increase in the percentage of short-circuits and by the decrease in the percentage of effective electrical spark discharges (refer to Fig. 5.10a).

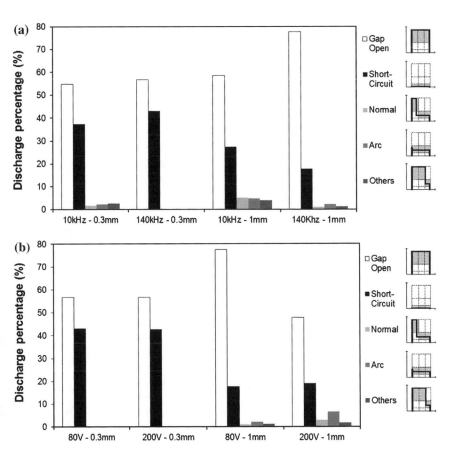

Fig. 5.10 a Influence of the frequency of electric current pulses and of, **b** open-circuit voltage in the type and percentage of electrical spark discharges (amplitude of current 1.5 A and diameters of the electrodes equal to 0.3 and 1 mm)

Data in Fig. 5.10 was automatically collected by the computer software that is capable of recognizing different types of discharge patterns (refer to Sect. 5.3.5 and Fig. 5.7). The increase in short-circuits for the electrodes with smaller diameters is attributed to its higher aspect ratio and to lack of rigidity, which create difficulties in keeping the gap and preventing the electrodes from shorting out against the workpiece as a result of vibrations induced by spark discharges and position devices (i.e., motion of the electrode holder). This will additionally create difficulties in the position control of the electrode.

Figure 5.9b allows concluding that MRRs of both electrodes increase with the increase in the open-circuit voltage and that MRRs of the electrodes with 1 mm diameter are up to 10 times higher than those obtained with electrodes of 0.3 mm diameter. A possible explanation for the increase in MMR with open-circuit voltage is related to the larger gap between the electrode and workpiece, which improves wash out by the flowing dielectric fluid and increases the overall efficiency of the electrical spark discharges. Larger gaps also facilitate the servo control to adjust the position of the Z-axis in order to obtain better and more efficient discharges.

5.4.4 Types of Electrical Spark Discharge

Figure 5.10a shows that the percentage of effective electrical spark discharges (i.e., normal, arc, and other sparks that efficiently remove material from the workpiece surface) diminishes as frequency increases for both 0.3 and 1 mm electrode diameters. This is attributed to smaller gaps between the electrode and workpiece and to difficulties in washing out debris and contaminants during shorter duration of the pulse off-times t_{off}.

The remaining discharges correspond to gap open- and short-circuits that are unable to create a plasma in the dielectric fluid. The percentage of occurrence of short-circuits is compatible with results presented by other researchers [12].

The percentage of the different types of effective electrical spark discharges as a function of the open-circuit voltage (Fig. 5.10b) shows a trend similar to that obtained with frequency. In fact, normal, arc, and other sparks that effectively remove material from the workpiece surface increase with open-circuit voltage in case of electrodes with 1 mm diameter whereas material removal mechanism for the electrodes with 0.3 mm diameter are basically due to, less effective, short-circuits.

5.4.5 Single Spark Discharges and Material Removal Mechanisms

The proposed μEDM prototype machine is capable of operating under single-electrical-spark-discharge mode. This hardware feature is not commonly found in commercial equipment and allowed investigating the influence of each type of sparking discharge in the material removal mechanisms and morphology of the craters.

The electrical spark discharges generated by the circuit shown in Fig. 5.5b ('impulse generator') under reverse polarity were analyzed by the computer program described in Sect. 5.3.5 and classified into three main groups (Fig. 5.11).

The first group corresponds to open circuits and therefore to absence of sparking discharge, plasma formation, and material removal. The second group includes three different types of electrical spark discharges (delayed, normal, and arc sparks), which effectively remove material from the surface of the workpiece. The third group includes complex discharges and short-circuits that are much less effective than those belonging to the second group in removing material from the surface of the workpiece.

The difference between the classification of sparks given in Fig. 5.11 and that available in one of the very few available research publications in the field [13] is the introduction of two additional types of sparks. The new sparks are named as 'delayed sparks' and 'complex sparks' and had been previously classified as 'other sparks' (refer to Fig. 5.10).

Delayed and complex sparks correspond to discharge patterns that were not automatically recognized by the aforementioned computer program (Sect. 5.3.5)

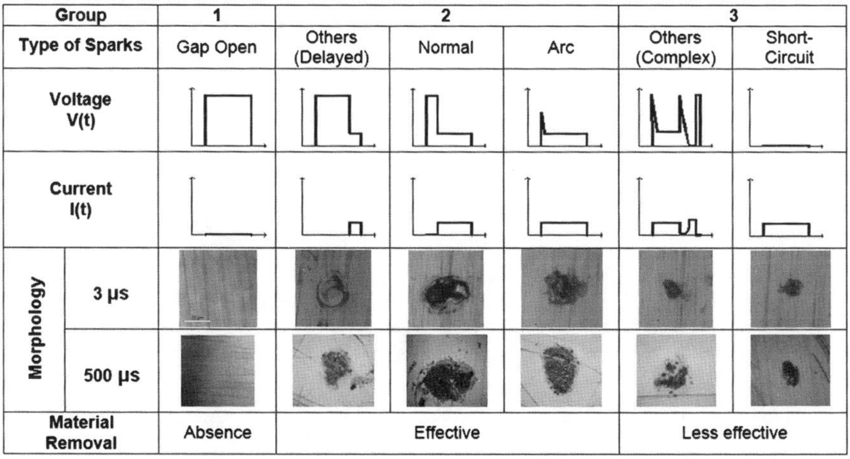

Group	1	2			3	
Type of Sparks	Gap Open	Others (Delayed)	Normal	Arc	Others (Complex)	Short-Circuit
Voltage V(t)						
Current I(t)						
Morphology 3 μs						
Morphology 500 μs						
Material Removal	Absence	Effective			Less effective	

Fig. 5.11 Classification of the different types of electrical spark discharges as a function of material removal and morphology of the craters. (single-electrical-spark-discharge mode with amplitude of current of 2 A and electrode diameter of 1 mm)

due to significant deviations from standard voltage versus pulse on-time evolutions. Delayed electrical sparks, for instance, present very large ionization times while complex electrical sparks show evidence of ionization and short-circuiting. Both types of sparks give rise to material removal but complex sparks are included in the third group for being less effective.

A possible reason for the lack of publications relating the type of electrical spark discharge with material removal mechanisms and morphology of craters can be the aforementioned limitations of commercial μEDMs to operate under single-spark mode. In fact, operation in single-electrical-spark-discharge mode allows investigating how duration of pulse on-time t_{on} influences the size of the craters (Figs. 5.12 and 5.13).

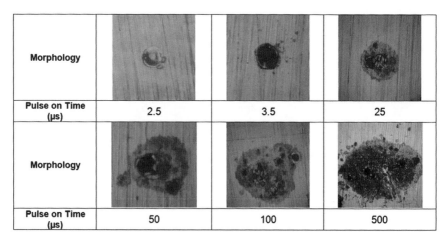

Morphology			
Pulse on Time (μs)	2.5	3.5	25
Morphology			
Pulse on Time (μs)	50	100	500

Fig. 5.12 Influence of pulse on-time on the morphology of the craters. (single-electrical-spark-discharge mode with an amplitude of current of 2 A and electrode diameter of 1 mm)

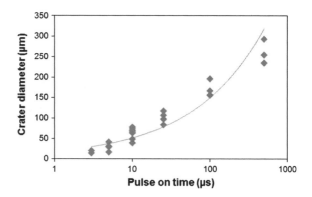

Fig. 5.13 Average diameter of craters as a function of the pulse on-time (open-circuit voltage 200 V with an amplitude of current of 2 A)

In general terms, the size of the craters increases as the duration of pulse on-time increases because the plasma channel formed in the dielectric enlarges and the energy is transmitted to a larger surface of the workpiece (Fig. 5.12). Because transmission to larger surfaces implies smaller energy densities, the resulting craters are less well defined than those generated with smaller duration of the pulse on-time (please refer to the left and rightmost pictures in Fig. 5.12).

The decrease in definition of the craters that is observed for larger pulse on-times t_{on} is attributed to the formation of multiple small craters during an electrical spark discharge as a consequence of secondary streamers that are formed at the vicinity of the main plasma channel. The result of multiple plasma streamers on the surface of the workpiece can be observed in Fig. 5.12 by referring to the photographs corresponding to pulse on-times equal and above 25 μs.

From an energy point of view, formation of secondary streamers is associated with instability of the primary plasma channel as pulse on-time increases, due to increasing of size and diminishing of energy density.

Figure 5.13 shows the average diameter of craters formed in single-electrical-spark-discharge mode as a function of the pulse on-time. There are two quite distinct trends: a leftmost trend where average diameter growths progressively with the pulse on-time and a subsequent trend where the rate of growth is exponential. The change in trend occurs for a pulse on-time around 10 μs and is physically attributed to the formation of multiple craters (over a larger surface of the workpiece) instead of single craters. Similar evolution of the average diameter of craters with the pulse on-time was reported by other researchers [14, 15].

The overall morphology of craters resulting from electrical spark discharges that produce single or multiple craters is shown in Fig. 5.14. The number of craters, geometry, area, aspect ratio (between crater depth and average diameter), borders, spatter, and formation of black layers are comprehensively systematized, and the work illustrates the potential of the μEDM prototype machine, operating under single-electrical-spark-discharge mode, for investigating material removal mechanisms.

The overall characterization of the morphology of the craters that are produced by a single electrical spark discharge (Fig. 5.14) is not commonly found in the research literature, and as far as authors are aware, only the works by Schulze et al. [16] and Rehbein et al. [17] briefly address this topic.

Although the investigation has been mainly performed with reverse polarity (that is, the workpiece connected to negative terminal of power), it was decided to check the combined effects of polarity and pulse on-time on the morphology of the craters. As shown in Fig. 5.15, direct polarity always produces a single, well-defined crater per electrical spark discharge whereas reverse polarity changes material removal mechanisms from single to multiple craters per electrical spark discharge when the pulse on-time increases.

The reason why direct polarity was not utilized throughout the investigation was due to larger electrode wear resulting from collisions of positive ions against the electrode surface and to lower stability of process control compared with that

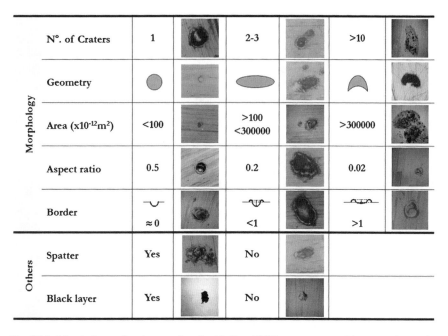

Fig. 5.14 Morphology of craters produced with the μEDM prototype machine operating under single-electrical-spark-discharge mode (normal type of sparks)

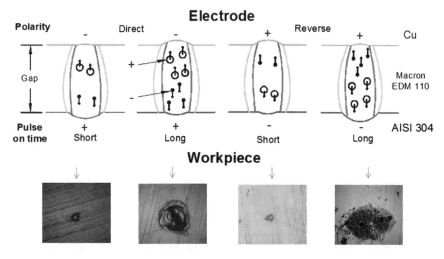

Fig. 5.15 Combined influence of polarity and pulse on-time on the morphology of craters produced with the μEDM prototype machine operating under single-electrical-spark-discharge mode (normal type of sparks)

obtained by reverse polarity. The flow of electrically charged particles is schematically depicted in Fig. 5.15.

5.4.6 Electrical Discharge Circuits and Material Removal Mechanisms

Figure 5.16 shows the influence of the three different interchangeable discharge circuits of the μEDM prototype machine (Fig. 5.5b–d) on the morphology of the craters.

Observation of the craters shown in Fig. 5.16 combined with measurements of their depth reveal that craters produced by the RC-type relaxation circuit present the highest ratios between depth and average diameter (Fig. 5.16b). This is because craters produced by the RC-type relaxation circuit are deeper than the others as a result of the discharge energy being instantly transferred without enough time for the plasma channel to increase its diameter and enlarge its working area on the workpiece surface.

The remaining two types of circuits produce similar craters (Fig. 5.16a–c). The influence of current flow with time, which is feasible to study with electronic pulse discharge circuit, is globally negligible although the overall morphology of the craters shows signs of slightly larger edges and spatters as well as black layer deposits.

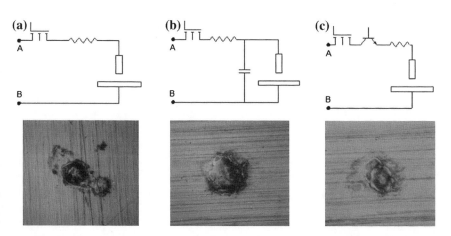

Fig. 5.16 Normal craters produced by the μEDM prototype machine operating under single-electrical-spark-discharge mode with **a** the impulse discharge circuit (open-circuit voltage 200 V, peak current 2 A, pulse on-time 10 μs), **b** the RC-type relaxation discharge circuit (open-circuit voltage 200 V, peak current 10 A, capacitor of 27nF), and **c** the electronic pulse discharge circuit (open-circuit voltage 200 V, peak current 2 A, pulse on-time 10 μs)

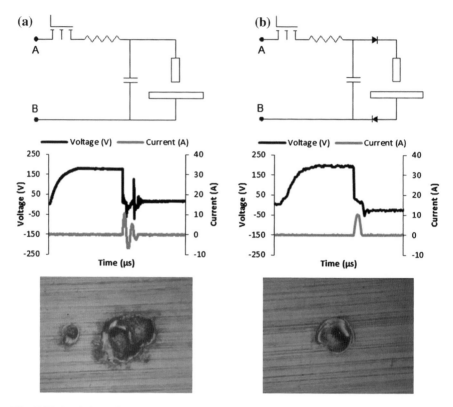

Fig. 5.17 Evolution of voltage and current with pulse on-time and photographs of craters obtained by the μEDM prototype machine operating under single-electrical-spark-discharge mode with (**a**) a conventional RC-type relaxation discharge circuit and (**b**) a modified RC-type relaxation circuit that includes two diodes

The utilization of RC-type relaxation circuits quite often leads to the formation of two craters (one much smaller than the other) during a single electrical spark discharge (Fig. 5.17a). This phenomenon is due to reverse of current flow between the capacitor and μEDM process and gives rise to a second plasma channel, which corresponds to negative peaks of voltage and current that are clearly seen in the evolutions of voltage and current with pulse on-time (Fig. 5.17a). Similar negative peaks of voltage and current with pulse on-time have been observed by other researchers [18].

By modifying the RC-type relaxation circuit to include two diodes that allow current to pass in one direction, while blocking current in the opposite (reverse) direction, there is only a plasma channel and only one crater (Fig. 5.17b). This additional control avoids reversing polarity during each sparking discharge and, therefore, helps diminishing the overall wear of the electrodes.

5.5 Conclusions

The chapter presents a prototype machine for μEDM capable of operating across a broad range of operative conditions that are suitable for educational and research purposes.

The design of the machine to incorporate easily interchangeable discharge circuits that quickly adapt the hardware to meet specific requirements of investigation allowed authors to test single- and multiple-electrical-spark-discharge mode and to better analyze and quantify the physical mechanisms behind material removal.

The computer program that automatically performs the recognition of the different types of electrical spark discharges that are generated during the μEDM process proved to be efficient in determining the percentage of the each type of discharge that occurred during the experiments. The program revealed crucial to investigate the physical mechanisms behind material removal.

Variations in the electrode diameter, frequency, and open-circuit voltage give rise to significant differences in MRR, effectiveness of electrical spark discharges and morphology of the craters. It was found that MRR increases with open-circuit voltage and that there is an optimum frequency that ensures the best MRR. This was attributed to competing effects of plasma efficiency, type of electrical spark discharges, adaptivity and reaction of the electrode position control, and effectiveness of flushing out debris and contaminants from the gap between the electrode and workpiece.

In addition, it was found that morphology of the craters is significantly influence by polarity and by the pulse on-time. In case of reverse polarity, multiple craters produced by secondary streamers of the plasma channel are likely to be observed when the pulse on-time is higher than 10 μs.

Acknowledgments Ivo Bragança and Gabriel Ribeiro would like to acknowledge Fundação para a Ciência e a Tecnologia of Portugal for providing the research grants SFRH/BD/48744/2008 and SFRH/BD/66576/2009. The authors would like to acknowledge the work of Paulo Gaspar.

References

1. Sommer C (2000) Non-traditional machining handbook. Advance Publishing, Houston
2. Uriarte L, Herrero A, Ivanov A, Oosterling H, Staemmler L, Tang PT, Allen D (2006) Comparison between microfabrication technologies for metal tooling. J Mech Eng Sci 220:1665–1676
3. Masuzawa T (2000) State of the Art of Micromachining. Ann CIRP 49:473–488
4. Rajurkar KP, Levy G, Malshe A, Sundaram MM, McGeough J, Hu X, Resnick R, DeSilva A (2006) Micro and nano machining by electro-physical and chemical processes. Ann CIRP 55:643–666
5. Khan DA, Hameedullah M (2011) Effect of tool polarity on the machining characteristics in electric discharge machining of silver steel and statistical modelling of the process. Int J Eng Sci Technol 3:5001–5010

6. Aligiri E, Yeo SH, Tan PC (2010) A new tool wear compensation method based on real-time estimation of material removal volume in micro-EDM. J Mater Process Technol 210:2292–2303
7. Peças P, Henriques E (2003) Influence of silicon powder-mixed dielectric on conventional electrical discharge machining. Int J Mach Tools Manuf 43:1465–1471
8. Yu S, Lee B, Lin W (2001) Waveform monitoring of electric discharge machining by wavelet transform. Int J Adv Manuf Technol 17:339–343
9. Schumacher BM (2004) After 60 years of EDM the discharge process remains still disputed. J Mater Process Technol 149:376–381
10. Kunieda M, Lauwers B, Rajurkar KP, Schumacher BM (2005) Advancing EDM through fundamental insight into the process. Ann CIRP 54:599–622
11. Ho K, Newman S (2003) State of the art electrical discharge machining (EDM). Int J Mach Tools Manuf 43:1287–1300
12. Joseph C (2005) Contribution à l'accroissement des performances du processus de μEDM par l'utilisation d'un robot à dynamique élevée et de haute précision. PhD Thesis, EPFL, Lausanne, Switzerland
13. Snoeys R, Dauw D, Jennes M (1982) Survey of EDM adaptive control and detection systems. Ann CIRP 32:483–489
14. Tosun N, Cogun C, Pihtili H (2003) The effect of cutting parameters on wire crater sizes in wire. Int J Adv Manuf Technol 21:857–865
15. Katz Z, Tibbles C (2007) Analysis of micro-scale EDM process. Int J Mach Tools Manuf 25:923–928
16. Schulze H, Hermsa R, Juhrb H, Schaetzinga W, Wollenberg G (2004) Comparison of measured and simulated crater morphology for EDM. J Mater Process Technol 149:316–322
17. Rehbein W, Schulze H, Mecke K, Wollenberg G, Storr M (2004) Influence of selected groups of additives on breakdown in EDM sinking. J Mater Process Technol 149(1–3):58–64
18. Öpöz TT (2008) Manufacturing of micro holes by using micro electric discharge machining (micro-EDM). MSc Thesis, Atilim University, Turkey

Chapter 6
Abrasive Water Jet Milling

Mukul Shukla

Abstract Abrasive water jet (AWJ) machining and abrasive water jet cutting (AWJC) are widely used, especially where very hard materials like titanium (Ti) alloys, high-strength steel, ceramics, etc. need to be machined or cut. In this chapter, an overview of the abrasive water jet milling (AWJM) process is presented. The essential challenge is at controlling the depth of cut (DoC) produced by varying the important AWJ machining process parameters. Experimental studies, process modeling and control based on FEM, artificial intelligence techniques and regression, and mechanisms of material removal are covered from the recent literature with the focus being on Titanium alloys. Experimental study and nonlinear regression–based process modeling of controlled depth AWJ milling of Grade 2 Ti alloy is also presented. Finally, various challenges including scope of future research in AWJM are highlighted.

6.1 Introduction

In recent times, numerous advanced machining processes have been introduced, giving better quality machining, while being competitively economical and efficient. These processes include those based on lasers, abrasive water jets (AWJ), plasma, and ion beams. AWJ milling which is a form of AWJ machining is gaining wide popularity owing to its speed and flexibility of processing a wide variety of materials. Newer grade of materials possessing superior properties, for example,

M. Shukla (✉)
Department of Mechanical Engineering Technology, University of Johannesburg,
South Africa, Johannesburg, South Africa
e-mail: mshukla@uj.ac.za; mukulshukla@mnnit.ac.in

M. Shukla
Department of Mechanical Engineering, MNNIT, Allahabad, UP, India

J. P. Davim (ed.), *Nontraditional Machining Processes*,
DOI: 10.1007/978-1-4471-5179-1_6, © Springer-Verlag London 2013

ceramics, composites, super alloys, are being used as alternatives to regular materials, for improved economy and increased quality and functionality. A typical machining application is of Titanium (Ti) and its alloys for aircraft components where the preference is for lighter materials with high strength.

6.1.1 Abrasive Water Jet Machining

Off late, cutting based on abrasive laden water jet and laser beam has proved to be a superior method than other traditional cutting methods. Abrasive water jet cutting (AWJC) is being widely used, especially in cutting of harder or low-machinability materials like Ti alloys, ceramics, metal matrix composites, concrete etc. An AWJ machine typically uses a multi-reciprocating pump as the primary energy source. Treated water is pumped to very high pressures in the range of 4,000–6,000 bar (400–600 MPa). An abrasive (e.g., garnet) is introduced into the water stream from an adjacent hopper and directed to a mixing chamber inside the cutting head. The abrasives are accelerated and exit the nozzle with the water through an orifice of small diameter (ranging from 0.1 to 1.0 mm) [1]. The water coming out of the orifice at high velocity (even beyond 1,000 m/s) is used for applications in cutting/machining of materials including toughened steel, ceramics, Kevlar fiber-reinforced polymers, and Ti among others, by the erosion process [2]. Figure 6.1 depicts the schematic of working of a typical AWJ machine.

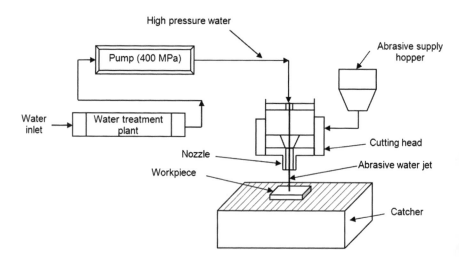

Fig. 6.1 Schematic of a typical AWJ machine

Some of the comparative advantages of AWJ machining over traditional material cutting methods include [3–5]:

1. Marginal thermal stresses or heat generation making it suitable for heat-sensitive materials like plastics.
2. Comparatively faster process.
3. Cut surface is smoother (Fig. 6.2) and requires limited post-processing.
4. Very thin pieces can be cut with least bending or melting.
5. Any contour can be cut (Fig. 6.2) on almost any material.

However, some of the disadvantages of the process are as follows: limited surface finish due to higher abrasive size, noisy, expensive equipment, unsafe if done manually, kerf taper and striated cut surface due to jet characteristics, used abrasives are an environmental hazard and moisture entrapment in workpiece.

Since abrasive water jet milling (AWJM)/C is a comparatively new process, many improvements and developments are ongoing to make the process more economical and standardized. One such application is targeted at AWJ controlled depth milling (CDM) of newer materials like Ti [7]. In-depth practical studies and process modeling for a wide variety of materials would greatly facilitate this implication.

6.1.1.1 Operations

The AWJ machining operations consist of cutting, multi-axis machining, milling, turning, drilling, polishing, etc. as listed below [8].

AWJ Cutting

Cutting is the most popularly used AWJ operation for various industrial applications. AWJ can be used for cutting of advanced materials, where it is required that the heat generated at tool tip should not pass into the workpiece, for maintaining

Fig. 6.2 A typical AWJM sample showing smooth cut contoured surfaces [6, 34]

structural integrity and other physical properties. However, the presence of surface striations and roughness present toward the bottom of the cut surface limit the applications of this technology.

AWJ Multi-Axis Machining

Multi-axis or 3D machining operations on flat objects using AWJs have been challenging owing to the incapability to monitor and control the depth of jet penetration. Two general approaches are generally used in practice for this: (1) using masks or templates to machine complex patterns like isogrid structures using selective pocketing [9] and (2) controlling the jet–workpiece interaction time through control of traverse rate, number of passes, and other process parameters [10]. Figure 6.3a and b, respectively, shows the multi-axis machining operation and multi-axis-machined carbon composite sample.

AWJ Milling

The AWJM occurs when traversing the jets with overlapping multiple passes across the workpiece surface. This multi-pass linear traverse cutting strategy utilizes the principle of superposition of several kerfs to obtain a defined geometry cavity. Several process parameters significantly contribute to the efficiency of the AWJM process as well as the final form of the generated cavities. AWJM of isogrid shapes is conducted to demonstrate the degree of control of milling depth (up to $0.001''$) attained using steel masks. Figure 6.4 shows the AWJ-milled slots in a Ti alloy workpiece.

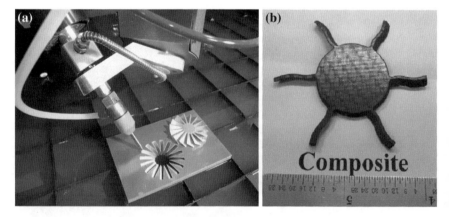

Fig. 6.3 a Multi-axis machining operation [wardjet.com] and b machined composite sample [11]

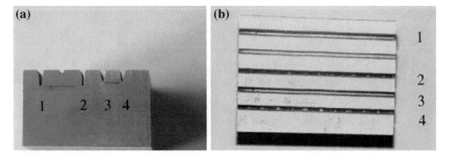

Fig. 6.4 a End view and **b** plan view of AWJ-milled slots in Ti alloy [12]

AWJ Turning

AWJ turning has been shown to be a suitable alternative process for turning or an additional preliminary rough turning technology to turn difficult-to-cut hypereutectic aluminum silicon or Ti aluminide alloys (for applications in the aerospace industry). Higher tool life combined with higher material removal rate (MRR), low process temperatures, and less modified material close to the cutting surface gives this cutting technology an edge over the conventional rough turning [13].

AWJ turning like conventional turning consists of spinning a specimen around the axis of rotation and simultaneously traversing the AWJ over the specimen in the desired contour to achieve an axisymmetric shape. The material removal takes place at the face of the workpiece rather than at the circumference at low traverse rates and vice versa. Figure 6.5a and b, respectively, shows the rough AWJ-turned Al specimen at 90° impact angle and with improved surface finish at small impact angle. Figure 6.6 shows the grooved and scalloped grinding wheel generated by the AWJ turning process.

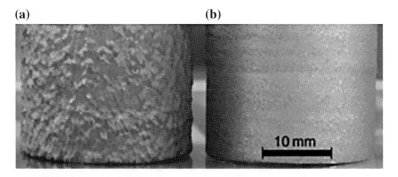

Fig. 6.5 AWJ-turned aluminum specimen at **a** normal and **b** small angle of impact [14]

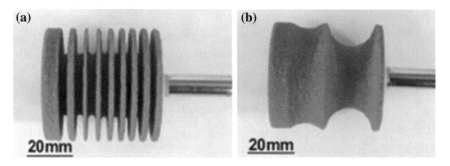

Fig. 6.6 AWJ-turned sample of **a** grooved and **b** scalloped grinding wheel [15]

AWJ Drilling

Hole generation by AWJ is realized by piercing, drilling, or trepanning operations. In piercing, neither the jet nor the specimen performs rotational movement. The jet just penetrates the material axially until it exits past the workpiece. The piercing process comprises of three phases: water jet impact, AWJ penetration, and AWJ dwelling. Size and tolerance of the pierced hole can be controlled by selecting suitable values of process parameters and controlling the dwell time. The longer the dwell, the larger the final hole size.

AWJ drilling is performed with a rotary water jet. The drilling rate linearly increases with an increase in pump pressure and abrasive mass flow rate. Holes can be drilled in different materials up to thickness of 15 cm, but the obtained surface roughness is comparatively low of the order of 2.2 µm.

Hole trepanning is a non-straight cutting process. However, due to the curvature of the cut, the geometry of the jet–material interface is more complex than that of a straight cut. During trepanning, due to the jet trail back, the diameter of the hole increases with increasing workpiece thickness, producing a poor quality hole [16]. The pattern of motion from the piercing location to the wall of the hole and accuracy of the traverse mechanism primarily control the roundness at the top side of the hole.

AWJ Polishing

In AWJ polishing, the sample rotates under the stationary jet and a ring of material is exposed to the jet effect. Here, the abrasives are injected externally in a high-speed hydrojet at shallow angles of attack in between the water jet and the workpiece under high pump pressures. AWJ can polish materials such as ceramics, stainless steel, and alloys. The polished surface quality strongly depends on the size and impact angle of the abrasive grains [8, 29]. A similar process is also used for the removal of coatings and scale.

6.1.1.2 Benefits

AWJ technology offers the following advantages in comparison with the conventional and other non-conventional techniques [7]:

High machining versatility: An abrasive water jet can cut through almost all ductile and brittle materials including many difficult-to-machine materials such as Ti alloys, high-strength advanced ceramics, high-strength steel, metal matrix, and Kevlar composites.

Negligible thermal distortion: The heat generated in AWJ process is instantaneously dissipated by the water. Thereby, there is hardly any temperature rise in the workpiece, leading to minimal changes in the material properties, microstructure, and structural integrity. This characteristic is especially useful for machining thermally sensitive materials such as metals, super alloys, and advanced ceramics.

Small cutting force and speedy setup: The cutting forces being very small, the chances of surface/subsurface damage to the cut material are minimal. Flat-surfaced samples can be directly positioned by laying them on a table without the need of any intricate clamping or tool changes.

Ability to generate contours: AWJs are exceptionally good at 2D machining. However, it is also possible to cut complicated 3D shapes or bevels of any angles and perform 3D profiling.

Eco-friendly: AWJC does not produce any harmful dust or particles that may pose a health hazard if inhaled and is considered to be one of the most environment-friendly machining process.

Availability of raw material: Water is used as the basic working fluid, and the abrasive materials most commonly used are garnet or silica, which are easily available at low cost.

However, AWJ machining also suffers from the following disadvantages:

- Overall the system is expensive in comparison with conventional techniques.
- Moisture entrapment issues;
- Noisy and unsafe;
- Recycling of abrasives.

6.1.1.3 Applications

Because of its technical performance and economics, AWJ machining is used in nearly all modern industries, such as automotive, aerospace, construction, mining, chemical process engineering (Table 6.1), and has numerous other potential applications too. It has been used particularly in pattern cutting of difficult-to-cut materials such as ceramics, laminated glass, and Ti sheets.

Table 6.1 Industrial applications of high-pressure water jets [7]

Industry	Applications
Civil engineering/construction	Cutting reinforced concrete, surface and joint cleaning, vibration-free demolition, soil stabilization and decontamination, water jet supported pile driving
Coal mining	Cutting metal structure, assist in drilling
Chemical and processing	Pipeline cleaning and coating, tube bundle cleaning, vessel, container, and autoclave cleaning
Maintenance and corrosion prevention	Coating removal, emission-free surface cleaning, selective paint sintering
Municipal engineering	Sewer cleaning
Automotive engineering	Deburring of parts
Environmental engineering	Material recycling, emission-free decontamination

6.1.2 Scope of the Chapter

This chapter is primarily limited to AWJ milling–based experimental studies, mechanisms of material removal, and process modeling and control (based on FEM, artificial intelligence techniques, and regression). The chapter mainly focuses on recent literature and AWJ machining of Ti alloys. Experimental study and nonlinear regression–based process modeling of controlled depth AWJ milling of Grade 2 Titanium alloy are presented. Finally, various challenges including scope of future research in controlled depth AWJM are also highlighted. The chapter does not cover the studies on abrasive jet machining, abrasive flow machining, and pure water jet machining.

6.2 Literature Review

AWJM takes place when an AWJ is used to remove material by erosion to a certain limited depth (i.e., not a through cut). AWJM is most feasible for materials that can be eroded more easily than cut (e.g., hard and/or brittle materials and certain tough fiber-reinforced (e.g., Kevlar/Aramid) polymers. AWJM involves the coordination between the bulk MRR and the proper overlap between successive kerf passes. Since the depth of cut (DoC) depends on numerous process parameters, controlling/restraining it while still maintaining desired surface finish has remained a challenge in AWJM applications.

Jet milling was first researched in 1987 by Hashish [17] where he investigated the volume removal rate and parameters that govern the surface topography. The shape of the cut/groove was mainly influenced by the stand-off distance and traverse rate, with higher traverse rates giving better surface topography. Hashish conducted another investigation to compare AWJ milling with traditional milling [18] and studied the effect of various parameters to obtain the best parameter combination that gives high MRR and acceptable surface finish.

A review of the current state of research and development in AWJC was presented in [8, 11, 19]. Another review of research articles of last 5 years on the application of evolutionary techniques in optimizing machining parameters was presented by [20].

6.2.1 Process Modeling and Mechanism of Material Removal

6.2.1.1 Straight Milling

Erosion modeling of AWJM of polycrystalline ceramics providing a relationship for the material effects of dense alumina ceramics on the process of erosion was presented in [21]. There exists a strong correlation between the erosion rate and the ratio of grain size/fracture energy. Experimental investigation of rectangular AWJ pocket milling of alumina ceramics was conducted by [22]. It was found that the depth per cycle decreases with an increase in traverse speed (TS) and reduction in abrasive flow rate (AFR). Experimental determination and feasibility study of using AWJs for CDM of aluminum and Ti isogrid parts used in aerospace and aircraft structures were published in [23]. TS was found to be the most critical parameter affecting the uniformity of milled surfaces and the operative mechanism of material removal. References [12, 24] applied AWJ for CDM of Ti6Al4V alloy. It was concluded that the surface waviness increases with number of passes of the jet over the workpiece.

In a study in AWJ-CDM of Ti alloy, it was found that MRR is high at high impingement angles (around 60°), while surface finish increased with an increase in impingement angle up to 45° [25]. It was demonstrated in [26] that in WJ pocket milling of Ti aluminide, the depth/pass increases with an increase in water jet pressure (WJP) and AFR. Another research study on AWJ-CDM of Ti6Al4V alloy considered particle hardness and shape and found that MRR for all the abrasives (brown aluminum oxide, white garnet grit, white aluminum oxide, and glass beads) is the highest at the lowest TS and decreases rapidly with increasing TS. Glass beads exhibit the lowest rates of the removal [27]. The effect of jet impingement angle and feed rate on the kerf geometry and dimensional characteristics in AWJM of 10-mm-thick SiC ceramic plate was investigated in [28]. They found that the kerf geometry is dependent on the variation in SOD, abrasive particle velocity distributions, and their local impact angles. The ductile erosion mechanism of hard–brittle materials by AWJ was investigated in [29]. Ductile erosion leads to micro-material removal and yields a smooth eroded surface without any fracture.

In another article, the identification and analysis of TS of cutting head in relation to Ti surface topography created by AWJC were performed [30]. Based on surface roughness, the abrasive water jet interaction, mechanism of stock removal, and a new classification of different qualitative zones were developed, including an often neglected initial zone. An optimal cutting head TS expression was also semi-empirically determined. New relations derived from quadratic sum of the tensile and pressure component of SIG_{def} used to predict a topographic function across the width of the cut were developed.

Based on the jet flow characteristics and erosion theories, Dadkhahipour et al. 2012 investigated the formation mechanisms of channels milled by AWJ on amorphous glass. It was found that the channels were formed through four different zones, i.e., an opening zone, a steady-cutting zone, an unsteady-cutting zone, and a finishing zone. These zones are respectively associated with a secondary viscous flow generated upon the jet impact on the top surface of material, a turbulent flow developed during the penetration of the jet into the material, a transition or laminar flow at the downstream of the jet, and a vortex and damping flow caused by the accumulation of the low-energy solid particles at the bottom of the channel. Bulges are found at the channel bottom and close to the channel wall machined at high nozzle speed as a result of a force induced by the acceleration/deceleration of the moving nozzle when changing direction during the operation. Sawtooth waves are generated on the machined surface for smaller cross feeds [31]. *A study of the micro-channeling process on amorphous glasses using an abrasive slurry jet is presented by* [32]. *The models account for different slurry and workpiece properties.*

The lack of methods for online monitoring of jet penetration (i.e., area of abraded footprint) makes it difficult to control the quality of the AWJM process. Rabani et al. [33] presented a method to control the jet penetration on AWJM, introducing a new concept based on transfer rate of energy (TRE). It links the input jet energy, area of abraded footprint, and jet feed velocity, exploiting its property to remain constant for a particular set of AFR and pump pressure. The input jet energy producing the part erosion is monitored using an acoustic emission (AE) sensor mounted on the workpiece surface, while the jet feed velocity is acquired online from the machine axis encoders. With the preevaluation of TRE as specific response to the set of AWJM parameters, the area of abraded jet footprint can be calculated online. Further, to make the method more powerful, the input jet energy has been related to the process operating parameters, while their constant values have been monitored via a pressure gauge and second AE sensor mounted on the focusing tube. The uniqueness of the proposed monitoring approach is based on the fact that TRE permits to know the amount of adjustments of the jet feed velocity required to keep the jet penetration constant in case any process disturbances occur. This monitoring methodology opens avenues for closed-loop control strategies of AWJM so that complex features can be generated with minimum human intervention.

In AWJM, a flat surface is obtained when the jet cuts single overlapping slots. The resulting surface depends not only on the process parameters but also on the lateral feed between adjacent slots. Many researchers demonstrated the capability of AWJ technology for precision milling operations in different materials using a mask to avoid problems related to the dynamic behavior of the machine. But the

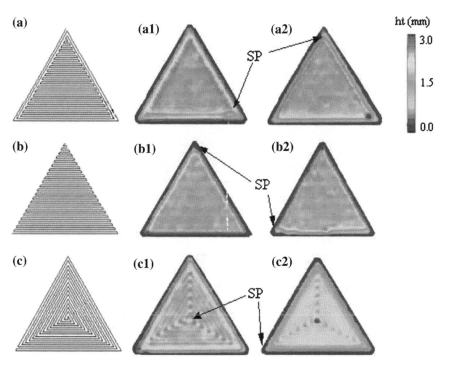

Fig. 6.7 Actual AWJ-milled surfaces for each starting point (SP). **a** Tool path #3, **b** tool path #4, and **c** tool path #5 [34]

limitation related to the use of mask is the additional cost and 2D. Alberdi et al. obtained an experimental model relating the total cavity depth with the depth and width of a single slot and with the lateral feed. Alternative tool paths were also studied aiming to find new strategies to allow maskless AWJ milling (Fig. 6.7) [34].

Another research focused on modeling of slot kerf profile produced by AWJM process consisting of overlapping single slots [35]. A hybrid evolutionary approach combining grammar-guided genetic programming (GGGP) together with genetic algorithms (GA) is proposed as automatic kerf profile model generator based on the maximum depth (h_{max}) and the full width at half-maximum (FWHM) (Fig. 6.8). Both h_{max} and FWHM were modeled using analysis of variance (ANOVA) and regression approach in terms of four important process parameters. The obtained expression after evolutionary system execution is a combination of two exponential functions (Eq. 6.1):

$$h(h_{max}, \text{FWHM}, r) = h_{max}\left(1.21e^{-\left(\frac{1.85r}{\text{FWHM}}\right)^2} - 0.208e^{-\left(\frac{4.45r}{\text{FWHM}}\right)^2}\right) - 12.2 \text{ [μm]} \quad (6.1)$$

CDM with AWJs is very difficult to conduct, due to the milled footprint's dependency on both the jet kinematics (e.g., TS or exposure time upon the workpiece and orientation of the jet relative to the target surface) and the jet

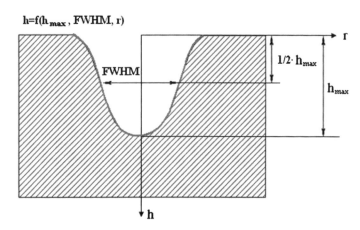

Fig. 6.8 Kerf shape characterization [35]

energy parameters (e.g., WJP, AFR). Anwar et al. [36] conducted modeling, simulation, and validation of AWJ footprint obtained in CDM at various jet TS and pump pressures at 90° angle of attack using the finite element method (FEM). The abrasive particles were modeled with various non-spherical shapes and sharp cutting edges, while the Ti6Al4V alloy extensively used in the aerospace industry is the workpiece material. The interaction between the AWJ plume and the target surface is accounted in the FEM material model by incorporating the effects of strain-rate sensitivity, adiabatic heating, and friction during the particles–workpiece interaction.

6.2.1.2 Multi-axis AWJ Milling

Multi-axis AWJ machining of cylindrical objects is relatively easy to perform by incorporating cutting, turning, and drilling in the same setup. However, 3D machining of flat objects using AWJ has always been a challenge. The capability to monitor and control the depth of penetration in AWJ cutting and drilling determines the effectiveness of multi-axis operations. Both analytical and empirical models have been developed for predicting the depth of penetration in 3D AWJ machining.

A robust model combining the particle kinematics and the constitutive equation for the particle erosion rate was developed. Based on it, a quantitative simulation of the 3D AWJ machining process was performed for local and global machining parameters such as the average DoC, surface roughness, and waviness. The validity of the 3D surface features generated through the model for different processes (such as drilling and cutting) was verified experimentally for glass, Ti, and other metals [37, 38]. Experiments were performed to test the application of 3D axis nozzle control for cutting 3D profile parts [39].

WJC and AWJC can be performed even at pressures up to 690 MPa [40] although normally available commercial systems are only capable of working at a maximum pressure of 379 MPa. Superior quality surfaces were produced at increased pressure, and the abrasive consumption was also significantly reduced. Reference [41] described the AWJC beyond the current industrial pressure limits. Firstly, the factors that limit the water pressure were discussed. Secondly, the jet formation was considered by addressing the effects of the geometry of the upstream tube and the orifice. Finally, the AWJC process was described in terms of energy transfer efficiency. There is an optimum abrasive load ratio over which the cutting ability of the jet decreases due to the less efficient power transfer from water jet to the abrasives.

A wide range of geometries that can be formed using multi-axis AWJ for different materials (carbon fiber composites used in aircraft fuselages to metal matrix composites) were studied by [11]. The influence of abrasive morphology and mechanical properties on workpiece grit embedment and cut quality in AWJC of a Ti alloy was investigated by [42]. Using profilometry, scanning electron microscopy (SEM), and energy dispersive X-ray analysis (EDX), grit embedment, surface waviness and roughness, and abrasive–surface interactions were evaluated. It was concluded that no significant variation in cut surface quality or abrasive particle embedment was observed in spite of using vastly different types of abrasives.

Kong et al. proposed a generic mathematical model applicable to different machine systems, with the benefit of simplicity by having fewer variables, for predicting maskless water-jetted footprints for arbitrarily moving jet paths [43]. This innovative footprint modeling approach has the key advantage of being independent of the properties of the workpiece material and/or machine setup, since it calibrates the specific etching rate. By considering any orientation of the jet plume vector relative to the target surface, this approach becomes a powerful tool for the development of advanced jet path strategies to enable AWJM of complex geometries.

A design of experiments (DOE) approach was taken, considering variables such as WJP, stand-off distance, TS, nozzle orifice diameter, AFR, and tool path step over distance in 3D AWJ pocket milling of Inconel 718. DoC and pocket geometry were the responses [44]. The results showed that WJP has a nonlinear behavior and is the most important variable for controlling the DoC and geometrical errors. Also, nozzle diameter and interaction between feed rate and abrasive mass flow are critical factors affecting the DoC.

6.2.2 DoC Modeling

Evans et al. presented an erosion model capable of predicting the DoC by impact damage in ceramics in the elastic–plastic response regime [45]. This model was adopted by [46] to obtain closed-form expression for the maximum DoC. Instead of using the typically used vertical cutting force monitoring which is costly and impractical, the online DoC monitoring based on the AE response was modeled [47]. It was established that AE is the most suitable technique for AWJ monitoring owing to its signal having high sensitivity to the DoC variation. A model for predicting the DoC in both oscillation and normal cutting mode was developed in [48]. Their experimental results showed that for ductile materials, nozzle oscillation during cutting at smaller angles and higher frequencies leads to efficient erosion process.

Other experimental DoC studies with controlled nozzle oscillation in AWJC of alumina ceramics were reported in [49, 50]. For the chosen cutting parameters, it was established that a high oscillation frequency (10–14 Hz) along with a low oscillation angle (4–6°) maximizes the DoC. Further, nozzle oscillation at small angles improved the DoC by up to 82 % on proper selection of cutting parameters. In [35], a model to predict the kerf profile (in terms of maximum cutting depth and width at the half of maximum modeled as a Gaussian function) in AWJ slot milling in aluminum 7075-T651 was introduced. The definition of an equivalent instantaneous traverse feed rate along the jet trajectory models the effect of jet acceleration. The model is capable of predicting the kerf profile at constant and variable traverse feed rate due to direction changing trajectories.

S. Harris in his article 'Abrasive water-jet model could enable lower-cost milling,' in The Engineer Online (Nov 25, 2011), highlights that new research could enable lower-cost milling on difficult materials using adapted water jet cutting machines [6]. Fluctuations in water pressure and geometry of abrasive particles mean that AWJM creates surfaces with variable depths. Engineers from research company Tecnalia and the University of the Basque Country in Spain have now developed a model for predicting and controlling the DoCs made by an AWJ machine, which was previously difficult to predict. They used information on process parameters as well as machine acceleration, tool path, and materials, to predict the depth at every point of machined surface (Fig. 6.9). This model is also used to find the optimum process parameters and milling strategy to reach a desired depth in any material and redesign AWJC machines to make them suitable for milling. However, in this report, there is no mention of the achievable surface finish, manufacturing of 3D forms, and comparison of machining time compared to conventional machining.

6.2.2.1 Simulation and Modeling

Experimentally, the eroded material volume loss may be measured and the erosion mechanism also investigated by analyzing the worn surface and erosion conditions. However, erosion is a complex phenomenon, governed by various process variables. Thus, it becomes practically very difficult to comprehend all this experimental information. Computer modeling allows 'virtual experiments' to be carried

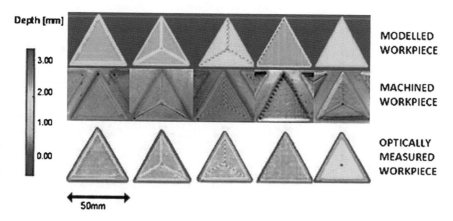

Fig. 6.9 Predicted and controlled DoCs made by an AWJ machine [6, 34]

out under controlled conditions and provides an effective method, complementary to experimental techniques, for a fundamental understanding of the process [51].

A novel approach was presented for modeling of 3D topography generated by AWJC [52]. The 2D topography at different depths of the cut surface was generated by considering the trajectories on the cutting front and the randomly impacting abrasive particles at the cut surface walls. Several 2D profiles generated in each region of cut were superimposed to obtain a net 3D topography. The nature of these 3D profiles was analyzed and validated using power spectral density analysis. In [53], theoretical and experimental studies to model the DoC based on the tilt angle of cutting head were conducted. The model was verified to be widely applicable on a variety of materials and established to be very reliable.

AI Techniques

Fuzzy set theory for selection of levels of the main AWJC parameters for a required DoC was applied in [54]. A neurogenetic approach to adaptively control the AWJC process to model the DoC and derive the optimum parameter settings by accounting for varying diameter of focusing nozzle was presented in [55]. Hybrid artificial neural network (ANN) and simulated annealing (SA) techniques were applied by Zain et al. [56] to estimate optimal process parameters considering a wide range of process parameters. They also integrated the soft computing techniques SA and GA to estimate optimal process parameters that lead to a minimum value of surface finish in AWJC [57]. The estimation of optimal solution using the integrated approach needs a smaller number of iterations compared to individual SA and GA optimization and yields improved machining performance in comparison with experiments and regression modeling. Vundavilli et al. [58] developed a fuzzy logic (FL)-based expert system for prediction of DoC in AWJ machining. It is important to note that the performance of the FL depends on its knowledge base. The following three FL approaches were used: (1) Mamdani based, (2) the database and rule base of the FL-system are optimized, and (3) the total FL-system is evolved automatically based on binary-coded GA. The accuracy predicted by the automatic FL-system is better than that of other two FL-systems. The expert system eliminates the need of elaborate experimentation, to select the most influential parameters affecting the DoC in AWJ machining.

FEM Based

Various authors have attempted to address the problem of modeling of material removal in erosion using numerical methods based on the finite element method (FEM). Numerical methods although are unable to provide a detailed microscopic insight into the cutting and/or plowing phenomenon of AWJM. Nevertheless, they are highly advantageous as they are capable of simulating the erosion or machining behavior under different conditions (type of material, particle speed, size, shape, impact angle, etc.), thus leading to significant cost reduction, as against the equivalent experimentation [51, 59]

Single Particle

AWJ machining was modeled using FEM and explaining the abrasive particle–workpiece interaction [60]. Single-particle impact modeling in AWJ machining was conducted by [61–63]. In their research, [64] attempted to predict the crater profile produced by single-particle impact for CDM using FEM, rather than first performing the simulation of full jet plume impingement. The main objective of this article was to simulate and experimentally validate the crater profile at different impact angles of abrasive particles in water jet for Ti-based Ti 6Al 4V superalloy.

Multi-Particle

A modified model was presented from Finnie's model for AWJ erosion. This modified model could even deal with curved surfaces and simulated multiple particle erosion [65]. Study of multi-particle solid particle erosion of metallic targets was reported by [66, 67]. Further references on single- and multi-particle erosion modeling using FEM are listed by the author [51]. They conducted FEM-based multi-particle (twenty) impact modeling for erosion in AWJ machining of Grade 5 Ti alloy. The influence of abrasive particle impact angle and size, and velocity on the crater sphericity and depth, and erosion rate has been investigated.

CFD Analysis

Liu conducted computational fluid dynamics (CFD) modeling for ultrahigh velocity AWJs using the commercial flow solver software, Fluent [68]. The dynamic characteristics of the jet flowing downstream from a fine nozzle were then simulated under both steady state and turbulent, and two- and three-phase flow conditions. The jet characteristics lead to a fundamental understanding of the kerf formation phenomenon in AWJC. Wang (2009) [49] also conducted a similar simulation study of the jet dynamic characteristics using CFD. Wang and Wang (2010) [69] conducted theoretical analysis and developed a two-fluid flow model based on the basic conservation principles. They analyzed quasi-two-dimensional flow field outside the nozzle used in AWJM. A control volume method based on a phase-coupled algorithm was used to solve the coupled pressure–velocity equations.

6.3 Materials and Methods

Ti Grade 2 one of the four unalloyed, commercially available pure Ti variants has the following composition: hydrogen 0.01 %, nitrogen < 0.03 %, carbon < 0.08 %, oxygen < 0.25 %, iron < 0.3 %, Ti balance [70]. It has almost similar properties to Grade 5 alloy (the most widely used Ti alloy) but has relatively low strength and is available at a significantly cheaper price. Some of the unique properties of Grade 2 Ti alloy include:

- Combination of strength, ductility, and toughness;
- Heat treatable, easy to fabricate and weld;
- Can be used at temperatures below 400 °C without loss of physical properties;
- Good tensile properties even at high temperatures.

Ti Grade 2 finds applications in aircraft structures and engine parts, prosthetic devices, turbine blades, marine hardware, desalination equipment, heat exchangers, condenser tubing, etc. [70].

The AWJ-CDM experiments were performed on an ultrahigh pressure, ultrahigh speed, computer numerically controlled, Flow MACH 4 4020b water jet cutting machine (Fig. 6.10). It is equipped with a dynamic XD (to cut straight without the taper effect) 5-axis cutting head (Fig. 6.11) and ultrapierce attachment (for assistance in hole piercing), making it one of the most modern machines available commercially [71]. The specifications of the machine are as shown in Table 6.2.

Various parameters affect the AWJC process. These include—TS of head, jet impact angle and diameter, operating WJP, orifice stand-off distance, AFR and properties (shape, size, hardness, etc.), number of jet passes, etc. [74]. Studying the full variety of AWJ parameters at different levels in a full factorial experimentation would require considerable time and material and is thus infeasible. A DOE-based approach was used in this study to ensure that a larger variety of process parameter combinations are studied with minimum number of experiments [75].

Fig. 6.10 A typical Flow MACH 4 4020b AWJ machine [72]

Fig. 6.11 Typical cutting head of Flow MACH 4 4020b AWJ machine [73]

The three most important parameters were selected based on available literature and machine processing limitations. The three parameters investigated in the present research along with their four levels are given in Table 6.3.

Table 6.2 Typical specifications of Flow MACH 4 4020b

Parameters	Specifications
Operating pressure	0–600 MPa (6,000 bar or 87,000 psi)
Orifice	0.381 mm
Nozzle	1.016 mm
Traverse speed	0–12.7 m/min
Table size	3 × 2 m
Power	50 HP

Table 6.3 Experimental parameter settings

Parameter	Levels
Water jet pressure (P_w)	100, 250, 400, 550 MPa
Traverse speed (U)	2.33, 3.25, 4.08, 5 mm/s
Abrasive mass flow rate (m_a)	0.317, 0.363, 0.408, 0.454 kg/min

During experimentation, the following parameters were kept at a constant value: number of jet passes—1, garnet abrasive size—#80 (180 μm), nozzle stand-off distance—5 mm, and angle of attack—90°. The Ti workpiece was in the form of a flat plate of dimensions 230 × 90 × 38 mm. A 2.5 mm constant distance was maintained between the consecutive slots/cuts of 15–40 mm length (Fig. 6.12). A full factorial experiment was performed with the three parameters set at four levels each making it a total of 64 experiments/trials. The experiments were replicated twice to account for errors if any. The DoC was measured using a digital Mitutoyo Vernier scale (least count = 0.1 mm). Owing to brevity, only partial experimental parameter settings and measured DoC values of the 2nd replicate, along with the regression model predicted and the % error between experimental- and model-predicted values, are included in Table 6.4.

6.4 Data Analysis and Modeling Results

Majority of the DoC models require prior information of the material's properties such as the material's flow stress which cannot be easily determined without experiments. Hence, the energy approach [76, 77] was used in the present work owing to its ease of use and generalization for any material.

$$h = K \frac{m_a^x P_w^y}{d_j \rho_w U^z} \tag{6.2}$$

where K is a constant, h is the DoC, m_a is the abrasive mass flow rate, P_w is the WJP, d_j is the jet diameter, ρ_w is the density of workpiece, and U is the jet TS. The constants of Eq. (6.2) were then derived for modeling using the nonlinear

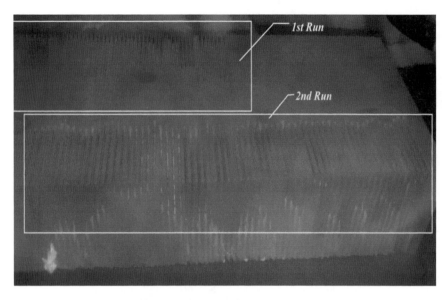

Fig. 6.12 Controlled depth AWJ-milled Ti workpiece [7]

regression software NLREG [78]. Based on the experimental results, the following nonlinear regression equation (Eq. 6.3) was generated (at 95 % confidence level):

$$h = 2.8109 \, \frac{m_a^{0.668381} P_w^{0.847245}}{d_j \rho_w U^{0.764906}} \tag{6.3}$$

From a comparison of coefficient of determination (R-sqr) values (Table 6.5), it can be observed that the first replicate's results did not give a very good fit with 86.69 % coefficient of determination (at a confidence level of 95 %). The second replicate, however, yielded improved results with a better coefficient of determination at 95.83 %. The mean coefficient of determination of the two runs was found to be 94.94 %. The percentage error is the ratio of the difference between the experimentally measured DoC and the regression model–predicted DoC to the experimental DoC. This is obtained from an average of all the trials in the two experimental runs. The maximum DoC deviation is the maximum difference between the experimentally measured DoC and the model-predicted DoC out of the full 64 cuts.

The average coefficient of determination was found to be 94.94 % which can be interpreted as the likeliness of obtaining similar results in future experiments or applications. This can be interpreted that if the above nonlinear model is used, the results or resulting DoC will have a 94.94 % likeliness of being as predicted. A linear regression model (Eq. 6.4) was also fitted from the experimental data to predict the DoC as follows:

$$h = 5.64 + 22.8 \, m_a + 0.0417 \, P_w - 3547 \, U \tag{6.4}$$

Table 6.4 Select experimental parameter settings, measured and estimated DoC

Run	m_a (kg/min)	P_w (MPa)	U (m/s)	DoC–2nd Run (mm)	Estimated DoC (mm)	% error
1	0.317	100	0.005	2.0	3.72	−86.11
2–4	–	–	–	–	–	–
5	0.317	250	0.00233	15.2	14.49	4.66
6	0.317	250	0.00325	12.4	11.25	9.30
7	0.317	250	0.00408	11.0	9.45	14.13
8	0.317	250	0.005	8.9	8.09	9.10
9	0.317	400	0.00233	23.2	21.58	6.98
10–14	–	–	–	–	–	–
15	0.317	550	0.00408	19.3	18.42	4.55
16	0.317	550	0.005	15.1	15.78	−4.49
17	0.363	100	0.00233	6.8	7.30	−7.35
18–30	–	–	–	–	–	–
31	0.363	550	0.00325	25.0	24.02	3.94
32	0.363	550	0.00233	31.6	30.94	2.08
33	0.408	550	0.00233	Through	33.46	–
34	0.408	550	0.00325	28.4	25.97	8.57
35	0.408	550	0.00408	22.0	21.81	0.88
36	0.408	550	0.005	20.4	18.68	8.44
37	0.408	400	0.005	14.3	14.26	0.27
38	0.408	400	0.00408	17.7	16.65	5.93
39	0.408	400	0.00325	20.0	19.83	0.87
40–52	–	–	–	–	–	–
53	0.454	250	0.005	10.0	10.29	−2.85
54	0.454	250	0.00408	14.0	12.01	14.22
55	0.454	250	0.00325	14.8	14.30	3.38
56	0.454	250	0.00233	19.7	18.42	6.48
57	0.454	400	0.00233	29.0	27.44	5.39
58–62	–	–	–	–	–	–
63	0.454	550	0.00325	28.9	27.89	3.50
64	0.454	550	0.00233	Through	35.93	–

Table 6.5 Statistical analysis of experimental data

Run	R-sqr (%)	% error	Max DoC deviation (mm)
1st run	86.69	35.5	11.7
2nd run	95.83	7.4	4.9
Average	94.94	9.04	4.9

The adequacy of the above regression model was also verified, and the coefficient of determination (R-sqr) was found to be 83 %. As expected in comparison with the nonlinear model, the linear model is unsuccessful to accurately model the DoC in AWJ-CDM.

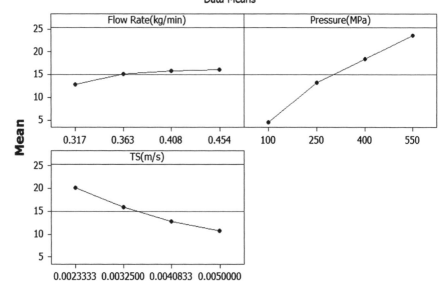

Fig. 6.13 Main effects plot for mean experimental DoC at different settings of the three AWJM process parameters

From Fig. 6.13 of the main effects plot (drawn using the Minitab statistical software [79]), it can be visualized that the WJP had the most dominant effect on DoC, followed by the TS and AFR. As the WJP and AFR increase, the DoC also increases. However, the impact of WJP on DoC is much more significant than AFR. Since the water pressure is the main source of energy in the water jet, an increase in this energy will mean that the jet has an increased eroding power which leads to an increased DoC. The decrease in DoC due to an increase in TS can be explained by the fact that the longer time is, the more the abrasives are exposed to the surface being cut, and deeper will be the DoC. Increasing the TS/feed rate means decreasing this jet-to-material exposure time and thus a decrease in the depth of the resulting cut. The slightly nonlinear graph of the three process parameters with the DoC suggests that they are related slightly nonlinearly and justifies the adequacy of the nonlinear model (Eq. 6.3) for DoC prediction.

Although the DoC results of the two replicates differed largely in few sporadic cases, the overall average % difference between the results from the two experiments was found to be 0.67 % which can be treated as fairly acceptable. Thus, the experimentation can be considered as being fairly repeatable. Further research investigations on the improvement of this aspect are still ongoing.

6.5 Conclusions and Future Research

The purpose of this chapter was to review the advances in the field of abrasive water jet milling and to develop an abrasive water jet milling DoC prediction model for Grade 2 Ti model. The ultimate idea is toward controlled depth cutting for milling of free-form shapes using AWJs.

Finally, the further challenges and scope of further research in the field of abrasive water jet milling are compiled in this section. This study cannot be complete if surface quality studies are not conducted simultaneously. For these results to be more useful, machining studies constrained with surface quality requirements must be conducted [80, 81]. Optimization and modeling studies to find the optimum operating parameter values for different materials with constrained surface finish requirements are also required [82, 83]. The models can be extended to include additional critical quality attributes such as surface roughness of samples, micro-structure, material embedment, etc. and process parameters such as abrasive type, abrasive grit, tool paths, etc. Influence of attributes such as particles' shape, rotation, multiple overlapping impacts, and effect of slurry jet hydrodynamics can be further investigated in AWJM.

To enhance the cutting performance, various new techniques have been proposed. These include forward angling the jet, controlled nozzle oscillation, and multi-pass cutting and need to be further explored. Studies on DoC using the advanced machines need to include the angle of attack as it has a significant impact for ductile materials. Modern machines can now cut at different angles (3D or multi-axis machines). Exploring the possibility to sculpt any free-form-shaped object is essential as done in conventional milling [84]. Alternative technologies to overcome the grit contamination problem need to be explored. Rabani–AWJ as a versatile effective machining technique requires an accurate controlled monitoring solution to diversify its capabilities throughout industries. However, little attempts have been made for fully automatic monitoring of AWJ milling so far to develop appropriate closed-loop control strategies for AWJ milling.

Many improvements and developments are ongoing to make the AWJ machining process more economical and standardized. In-depth experimental studies combined with process modeling for a wide range of materials would greatly facilitate this mission.

Acknowledgments The financial assistance provided by the Faculty of Engineering and Built Environment, University of Johannesburg, in conducting the experimental studies is greatly acknowledged. Thanks are also due to Dr T U Siddiqui, Dr P B Tambe, Mr Naresh Kumar, and Mr R Gudani, my research students, and to Mr Deon of SA Stainless, Johannesburg, for allowing to conduct experiments on his Flow AWJ machining center.

References

1. WaterJets (2012) www.waterjets.org. Last accessed December 2012
2. Developments in Abrasive WaterJet Technology (2012) http://www.wjta.org/wjta/New_ Developments_etc.asp. Last accessed December 2012
3. WaterJet Machining (2012) http://www.nottingham.ac.uk/nimrc/research/advanced manufacturing/waterjet-machining.aspx. Last accessed December 2012
4. Fowler G (2003) Abrasive water-jet-controlled depth milling of titanium alloys. PhD Thesis, Nottingham University, pp 4–56
5. Hashish M (1987) Conference on wear of materials. In: Proceedings of Internet Texas, ASME, NY, pp 769–776
6. http://www.theengineer.co.uk/production-engineering/news/abrasive-water-jet-model-could-enable-lower-cost-milling/1011031.article#ixzz2EUuX9zVB. Last accessed December 2012
7. Gudani R, Shukla M (2012) Controlled depth abrasive water jet cutting of grade 2 titanium and regression modeling. Int J Mech Eng Mater Sci 5(2):117–122
8. Siddiqui TU (2010) Abrasive water jet cutting of continuous fiber-reinforced polymer composites: experimental studies, modeling and multi-objective optimization. Unpublished PhD thesis, MNNIT Allahabad
9. Hashish M (1994) Three-dimensional machining with abrasive waterjets, waterjet cutting technology. Mechanical Engineering Publications, Ltd, London, pp 605–633
10. Kovacevic R, Hashish M, Mohan R, Ramulu M, Kim TJ, Geskin ES (1997) State of the art of research and development in abrasive waterjet machining. Trans ASME 119:776–785
11. Folkes J (2009) Waterjet–an innovative tool for manufacturing. J Mater Process Technol 209(20):6181–6189
12. Fowler G, Shipway PH, Pashby IR (2005) Abrasive water-jet controlled depth milling of Ti6Al4V alloy—an investigation of the role of jet–workpiece traverse speed and abrasive grit size on the characteristics of the milled material. J Mater Process Technol 161:407–414
13. Uhlmann E, Flögel K, Kretzschmar M, Faltin F (2012) Abrasive waterjet turning of high performance materials. In: 5th CIRP conference on high performance cutting 2012, Procedia CIRP 1, pp 409–413
14. Manu R, Ramesh Babu N (2008) Influence of jet impact angle on part geometry in abrasive waterjet turning of aluminium alloys. Int J Mach Mach Mater 3(1/2):120–132
15. Axinte DA, Stepanian JP, Kong MC, McGourlay J (2009) Abrasive waterjet turning—an efficient method to profile and dress grinding wheels. Int J Mach Tools Manuf 49(3–4):351–356
16. Siddiqui TU, Shukla M (2011) Abrasive waterjet hole trepanning of thick Kevlar-epoxy composites for ballistic applications–experimental investigations and analysis using design of experiments methodology. Int J Mach Mach Mater 10(3):172–186
17. Hashish M (1987) Turning with abrasive waterjets—a first investigation. ASME J Eng Indus 109(4):281–290
18. Hashish M (1991) Characteristics of surfaces machined with abrasive waterjets. J Eng Mater Technol Trans ASME 113(3):354–362
19. Selvan MCP, Raju NMS (2011) Review of the current state of research and development in abrasive waterjet cutting. Int J Adv Eng Sci Technol 11(2):267–275
20. N. Yusup, Zain AM, Hashim SZM (2012) Evolutionary techniques in optimizing machining parameters: review and recent applications (2007–2011). Expert Syst Appl 39:9909–9927
21. Zeng J, Kim TJ (1996) An erosion model of polycrystalline ceramics in abrasive waterjet cutting. Wear 193(2):207–217
22. Paul S, Hoogstrate AM, van Luttervelt CA, Kals HJJ (1998) An experimental investigation of rectangular pocket milling with abrasive water jet. J Mater Process Technol 73:179–188
23. Hashish M, Monserud D (1990) Abrasive waterjet machining of isogrid structures. Quest Integrated Inc., Report QUEST TR-508, pp 63

24. Shipway PH, Fowler G, Pashby IR (2005) Characteristics of the surface of a titanium alloy following milling with abrasive waterjets. Wear 258:123–132
25. Fowler G, Shipway PH, Pashby IR (2008) An investigation of the role of jet impingement angle on process efficiency and surface characteristics for abrasive waterjet milling of Ti6Al4V. In: Proceedings of the 19th international conference on water jetting, Nottingham, UK, pp 353–364
26. Hashish M (2008) Waterjet pocket milling of titanium aluminide. In: Proceedings of the 19th international conference on water jetting, Nottingham, UK, pp 365–376
27. Fowler G, Pashby IR, Shipway PH (2009) The effect of particle hardness and shape when abrasive water jet milling titanium alloy Ti6Al4V. Wear 266:613–620
28. Srinivasu DS, Axinte DA, Shipway PH, Folkes J (2009) Influence of kinematic operating parameters on kerf geometry in abrasive waterjet machining of silicon carbide ceramics. Int J Mach Tools Manuf 49:1077–1088
29. Zhu HT, Huang CZ, Wang J, Li QL, Che CL (2009) Experimental study on abrasive waterjet polishing for hard-brittle materials. Int J Mach Tools Manuf 49(7–8):569–578
30. Hloch S, Valicek J (2011) Prediction of distribution relationship of titanium surface topography created by abrasive waterjet. Int J Surf Sci Eng 5(2/3)
31. Dadkhahipour K, Nguyen T, Wang J (2012) Mechanisms of channel formation on glasses by abrasive waterjet milling. Wear 292–293:1–10
32. Pang KL, Nguyen T, Fan JM, Wang J (2012) Modelling of the micro-channelling process on glasses using an abrasive slurry jet. Int J Mach Tools Manuf 53:118–126
33. Rabani A, Marinescu I, Axinte D (2012) Acoustic emission energy transfer rate: a method for monitoring abrasive waterjet milling. Int J Mach Tools Manuf 61:80–89
34. Alberdi A, Rivero A, de Lacalle LNL (2011) Experimental study of the slot overlapping and tool path variation effect in abrasive waterjet milling. J Manuf Sci Eng 133(3):034502
35. Alberdi A, Rivero A, Carrascal A, Lamikiz A (2012) Kerf profile modelling in abrasive waterjet milling. Mater Sci Forum 713:91–96
36. Anwar S, Axinte DA, Becker AA (2013) Finite element modelling of abrasive waterjet milled footprints. J Mater Process Technol 213:180–193
37. Kovacevic R, Yong Z (1996) Modeling of 3D abrasive waterjet machining, part I: theoretical basis, jetting technology. Institution of Mechanical Engineers, pp 73–82
38. Yong Z, Kovacevic R (1996) Modeling of 3D abrasive waterjet machining, part II: simulation of machining, jetting technology. Institution of Mechanical Engineers, pp 83–89
39. Duflou JR, Kruth JP, Bohez EL (2001) Contour cutting of pre-formed parts with abrasive waterjet using 3-axis nozzle control. J Mater Process Technol 115(1):38–43
40. Hashish M (2005) Economics of abrasive-waterjet cutting at 600 MPA pressure. In: Proceedings of WJTA American waterjet conference, Houston, Texas, Paper 4A-3, pp 1–14
41. Hoogstrate AM, Susuzlu T, Karpuschewski B (2006) High performance cutting with abrasive waterjets beyond 400 MPa. CIRP Ann Manuf Technol 55(1):1–4
42. Boud F, Carpenter C, Folkes J, Shipway PH (2010) Abrasive waterjet cutting of a titanium alloy: the influence of abrasive morphology and mechanical properties on workpiece grit embedment and cut quality. J Mater Process Technol 210(15):2197–2205
43. Kong MC, Anwar S, Billingham J, Axinte DA (2012) Mathematical modeling of abrasive waterjet footprints for arbitrarily moving jets: partI—single straight paths. Int J Mach Tools Manuf 53:58–68
44. Palafox GAE, Gault RS, Ridgway K (2012) Characterisation of abrasive water-jet process for pocket milling in Inconel 718. In: 5th CIRP conference on high performance cutting, procedia CIRP 1 (2012), pp 404–408
45. Evans AG, Gulden ME, Rosenblatt ME (1978) Impact damage in brittle materials in the elastic-plastic response regime. Proc R Soc Lon A 361:343–365
46. Abdel-Rahman AA, El-Domiaty AA (1998) Maximum depth of cut for ceramics using abrasive waterjet technique. Wear 218(2):216–222
47. Hassan A, Chen C, Kovacevic R (2004) On-line monitoring of depth of cut in AWJ cutting. Int J Mach Tools Manuf 44:595–605

48. Lemma E, Deam R, Chen L (2005) Maximum depth of cut and mechanics of erosion in AWJ oscillation cutting of ductile materials. J Mater Process Technol 160(2):188–197
49. Wang J (2007) Predictive depth of jet penetration models for abrasive waterjet cutting of alumina ceramics. Int J Mech Sci 49(3):306–316
50. Wang J (2009) A new model for predicting the depth of cut in abrasive waterjet contouring of alumina ceramics. J Mater Process Technol 209(5):2314–2320
51. Kumar N, Shukla M (2012) Finite element analysis of multi-particle impact on erosion in abrasive water jet machining of titanium alloy. J Comput Appl Math 236(18):4600–4610
52. Vikram G, Ramesh Babu N (2002) Modelling and analysis of abrasive waterjet cut surface topography. Int J Mach Tools Manuf 42:1345–1354
53. Hlavac LM (2009) Investigation of the abrasive water jet trajectory curvature inside the kerf. J Mater Process Technol 209(8):4154–4161
54. Kovacevic R, Fang M (1994) Modeling of the influence of the abrasive waterjet cutting parameters on the depth of cut based on fuzzy rules. Int J Mach Tools Manuf 34(1):55–72
55. Srinivasu DS, Ramesh Babu N (2008) A neuro-genetic approach for selection of process parameters in abrasive waterjet cutting considering variation in diameter of focusing nozzle. Appl Soft Comput 8(1):809–819
56. Zain AM, Haron H, Sharif S (2011) Estimation of the minimum machining performance in the abrasive waterjet machining using integrated ANN-SA. Expert Syst Appl 38(7):8316–8326
57. Zain AM, Haron H, Sharif S (2011) Optimization of process parameters in the abrasive waterjet machining using integrated SA–GA. Appl Soft Comput 11:5350–5359
58. Vundavilli PR, Parappagoudar MB, Kodali SP, Benguluri S (2012) Fuzzy logic-based expert system for prediction of depth of cut in abrasive water jet machining process. Knowledge-Based Systems 27:456–464
59. Kumar N, Shukla M, Patel RK (2010) Finite element modeling of erosive wear in abrasive jet machining. In: International conference on theoretical, applied, computational and experimental mechanics, ICTACEM, IIT Kharagpur, India, Paper 168
60. Hassan AI, Kosmol J (2000) Finite element modeling of abrasive water-jet machining. In: Proceedings of the 15th International conference on jetting technology. Ronneby (Sweden): BHR Group, pp 321–33
61. Junkar M, Jurisevic B, Fajdiga M, Grah M (2006) Finite element analysis of single-particle impact in abrasive water jet machining. Int J Impact Eng 32:7
62. Ahmadi-Brooghani SY, Hassanzadeh H, Kahhal P (2007) Modeling of single-particle impact in abrasive water jet machining. Int J Mech Sys Sci Eng 1:4
63. Takaffoli M, Papini M (2009) Finite element analysis of single impacts of angular particles on ductile targets. Wear 267:144–151
64. Anwar S, Axinte DA, Becker AA (2011) Finite element modelling of a single-particle impact during abrasive waterjet milling. In: Proceedings of the Institution of Mechanical Engineers, part J: Journal of Engineering Tribology, August 2011, vol 225, 8, pp 821–832
65. ElTobgy MS, Ng E, Elbestawi MA (2005) Finite element modeling of erosive wear. Int J Mach Tools Manuf 45:1337–1346
66. Molinari JF, Ortiz M (2002) A study of solid-particle erosion of metallic targets. Impact Eng 27:347–358
67. Shimizu K, Noguchi T, Seitoh H, Okadab M, Matsubara Y (2001) FEM analysis of erosive wear. Wear 250:779–784
68. Liu H http://www.eprints.qut.edu.au/16110/1/Hua_Liu_Thesis.pdf
69. Wang R, Wang M (2010) A two-fluid model of abrasive waterjet. J Mater Process Technol 210(1):190–196
70. Arcam Titanium Grade 2 (2012) www.arcam.com/CommonResources/Files/arcam.com/Documents/EBM%20Materials/Arcam-Titanium-Grade-2.pdf. Last accessed December 2012
71. Flow Mach 4 AWJ machines (2012) http://www.flowwaterjet.com/en/waterjet-cutting/cutting-systems/mach-4.aspx. Last accessed December 2012
72. www.precisionmachinerysales.com/waterjet/1392.htm. Last accessed December 2012

73. http://www.sawaterjet.co.za/photo_gallery/index2.html. Last accessed December 2012
74. Momber AW, Kovacevic R (1998) Principles of abrasive water jet machining. Springer, London
75. Montgomery DC (2001) Design and analysis of experiments, 5th edn. Oxford Publications, New York
76. Wang J (2007) Predictive depth of jet penetration models for abrasive waterjet cutting of alumina ceramics. Int J Mech Sci 49:306–316
77. Siddiqui TU, Shukla M (2010) Modeling of depth of cut in abrasive waterjet cutting of thick kevlar-epoxy composites. Key Eng Mater 443:423–427
78. NLREG (2012) www.nlreg.com. Last accessed December 2012
79. Minitab (2012) www.minitab.com. Last accessed December 2012
80. Alberdi A, Rivero A, López de Lacalle LN, Suárez A (2010) Effect of process parameter on the kerf geometry in abrasive water jet milling. Int J Adv Manuf Technol 51:467–480
81. Shukla M, Tambe PB (2013) Genetic algorithm based optimization of material removal rate with surface finish constraints in abrasive water jet cutting of carbon-epoxy composites. Accepted in Natural Computing
82. Siddiqui TU, Shukla M (2012) Modeling and optimization of abrasive water jet cutting of kevlar fiber-reinforced polymer composites, in "computational methods for optimizing manufacturing technology—models and techniques". IGI Global, USA, pp 262–286
83. Shukla M, Tambe PB (2010) Predictive modeling of surface roughness and kerf widths in abrasive water jet cutting of kevlar composites using neural network. Int J Mach Mach Mater 8(1 & 2):226–246
84. Borkowski J (2010) Application of abrasive-water jet technology for material sculpturing. Trans Can Soc Mech Eng 34(3–4):389–398

Chapter 7
A New Approach for the Production of Blades by Hybrid Processes

A. Calleja, A. Fernández, A. Rodriguez, L. N. López de Lacalle and A. Lamikiz

Abstract Manufacturing complex surfaces for high responsibility rotary components is critical in several applications, such as blades for turbomachinery, forged crankpins, iron casting crankshafts, grooved features of long power shafts and others. Nowadays there is a trend toward the performance of all required operations in the same machine tool, following the so-called multitasking approach. Traditional machine tools kinematic has developed into a new concept of multitasking machine, and even other non traditional processes such as laser tempering, laser cladding, rolling or burnishing are performed in the same machine and workpiece setup.In this chapter a new approach for the production of blades is deeply explained. A hybrid process for blades manufacturing is performed in a multitasking machine, by successive application of three no usual processes: turn-milling for cylindrical part definition, turn-ball-milling for contouring free form surfaces and finally burnishing for improving both final roughness and fatigue endurance. The outcome of this new approach is the obtaintion of functional complex parts with high integrity, quality and shortened lead times.

7.1 Introduction

The book now in reader's hands is focused on non-traditional machining processes for complex surfaces production regarding different applications. The manufacturing process of blades for turbomachinery joins together the use of several non-traditional machining processes. All these hybrid processes developed in the same multitasking machine lead to the obtainment of the final part. High-speed ball-end milling is the most spread technology currently used for free-form or sculptured

A. Calleja (✉) · A. Fernández · A. Rodriguez · L. N. López de Lacalle · A. Lamikiz
Department of Mechanical Engineering, Faculty of Engineering of Bilbao,
University of the Basque Country, C/Alameda de Urquijo s/n 48013 Bilbao, Spain
e-mail: amaia.calleja@ehu.es

J. P. Davim (ed.), *Nontraditional Machining Processes*,
DOI: 10.1007/978-1-4471-5179-1_7, © Springer-Verlag London 2013

surfaces machining [1]. This chapter is devoted to describe the production of this kind of pieces.

In this field, the development of cutting tools, machining strategies, CAM software and machine's programming are pieces of the same solution [2–4].

7.1.1 Blade-Ruled Surfaces and Applications

Blade parts are a common geometry in many sectors such as aeronautics and energy. They are made of complex surfaces, and they are subjected to extreme working conditions. For a proper manufacturing process, the aerodynamics of the part and the forces and stresses acting on the working blade have to be taken into account. The demanding manufacturing process of these parts needs to meet all the requirements to obtain the final part.

7.1.2 Propeller Blade Aerodynamics

The airplane propeller consists of two or more blades and a central hub to which the blades are attached. Each blade of an airplane propeller is essentially a rotating wing. As a result of their construction, the propeller blades are like airfoils, Fig. 7.1, and produce forces that create the thrust to pull, or push, the airplane through the air [5].

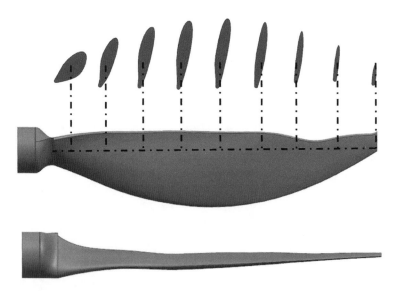

Fig. 7.1 Airfoil sections of a propeller blade

7.1.3 Forces and Stresses Acting on a Propeller Blade

The power needed to rotate the propeller blades is furnished by the engine. The engine rotates the airfoils of the blades through the air at high speeds, and the propeller transforms the rotary power of the engine into forward thrust [6]. An airplane moving through the air creates a drag force opposing its forward motion. Consequently, for an airplane to be able to fly, there must be a force applied to it that is equal to the drag, but acting forward. This force is called "thrust".

As may be seen in Fig. 7.2, the forces acting on a propeller in flight are as follows:

- **Thrust** is the air force on the propeller which is parallel to the direction of advance and induces bending stress in the propeller.
- **Centrifugal force** is caused by rotation of the propeller and tends to throw the blade out from the centre.
- **Torsion or twisting forces** in the blade itself caused by the resultant of air forces which tend to twist the blades towards a lower blade angle.

The stress acting on a propeller in flight is as follows:

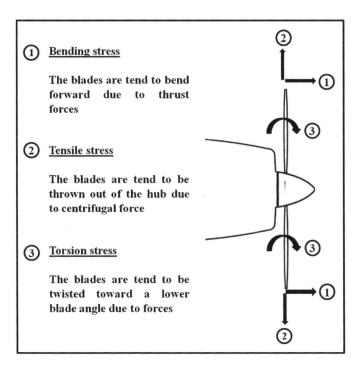

① Bending stress

The blades are tend to bend forward due to thrust forces

② Tensile stress

The blades are tend to be thrown out of the hub due to centrifugal force

③ Torsion stress

The blades are tend to be twisted toward a lower blade angle due to forces

Fig. 7.2 The stress acting on a propeller blade

1. **Bending stresses** are induced by the trust forces. These stresses tend to bend the blade forward as the airplane is moved through the air by the propeller.
2. **Tensile stresses** are caused by centrifugal force.
3. **Torsion stresses** are produced in rotating propeller blades by two twisting moments. One of these stresses is caused by the air reaction on the blades and is called the aerodynamic twisting moment. The other stress is caused by centrifugal force and is called the centrifugal twisting moment.

7.2 Machine Tool Sector Role on Non-Traditional Machining Processes

The industrial sector devoted to design and manufacture of machine tools is oriented to the conception of new products applied to their machines to be able to perform the maximal number of the workpiece operations. This way, it is possible to avoid workpiece multiple fixtures, being unnecessary to move the part from one position to another or moving it into a different machine. This new machine tool concept joining together as many manufacturing operations as possible is defined as "Multitasking machine". The use of these multitasking machines offers many advantages, both economic and manufacturing advantages. From the economical point of view, the initial inversion realized when buying the multitasking machine is easily recovered. The production is fastened because the number of operations is enlarged and at the same time fixturing times are reduced. Better manufacturing times are mainly shown in the reduction or absence of part movements to other machines. From the manufacturing point of view, the obtainment of the required part design tolerances is easily obtained due to the fixtures reduction, it is not necessary to take the part off, and this way more machining operations are performed with the same machining reference. In relation to the machining tolerances, more precision is gained because the final precision obtained is the one given by a machine and not the one given by several machines.

The multitasking machine is defined as a machine able to mill and to turn different prismatic forms [7]. In order to build a multitasking machine, it is possible to build it from the structure of a turning machine plus three different aspects: Y-axis perpendicular to X–Z plane, control of turning C-axis, which allows the part orientation, and live turning tools located in the bottom turret or in the B-axis head. The B-axis head should be able to be blocked in a concrete position to maintain a turning tool in a concrete position for conventional turning operations. If some of these aspects are not fulfilled, the machine will be a turning centre but not a multitasking centre. It is also possible to build a multitasking machine from a 4-axis milling centre, three linear axes and one rotary axis where the machine base is replaced and a turning machine base is added. If the rotary axis cannot move the plate up to the requiring turning plate speed or if the added torque is not enough, then the machine is just a milling centre and not a multitasking machine.

The major advantages developed by this technology in the last years are shown below. They represent the main evolution occurred in the machine tool sector.

- Year 1982: WFL machine tool building company creates WNC 500S MILL-TURN. A 4-axis turning machine capable of 5-axis continuous milling. This machine is part of a larger machine family where WNC 700S machine, presented in the 1984 Metav showroom, outstands.
- Year 1999: Mazak opens the door to the expansion of this technology presenting the INTEGREX machine in the EMO showroom.
- Year 2000: Nakamura® Tome company launches the Super Multitasking Machine "STW-40".
- Year 2004: Mori Seiki® starts the development of the NT series and the Stama concept integrated in their turning-based machines.
- Year 2005: Vertical turning machines introduce bigger functionality to milling spindles; this is where the turn-milling idea comes from.
- Year 2007: GMTK® designs their revolutionary concept including a turning and a milling centre. The design is based on the client requirements, according to the Mori Seiki® and Mazak® philosophy concept that has already set a point apart the traditional clichés. GMTK® philosophy resumes their new saying: *An objective above the ways*, that is, the part is solved with a multitasking machine.
- Year 2008: The multitasking revolution 2.0 starts both machines and tools are designed following the client requirements. New concepts such as the Sandvik multiturrets and milling heads with increasing power and revolutions appear.
- Year 2009: Spanish machine tool builders: GMTK®, Etxetar®, Ibarmia®, Danobat®, CMZ®, etc., have also developed and take part in the multitasking revolution. They are also supported by UPV/EHU and technological centres: Ideko, Fatronik, Tekniker, CIM.

The multitasking concept and the increasing interpolated axis included in new machine tools have generated new machining possibilities, as can be seen in Fig. 7.3. Turn-milling is a process in which the part rotation and the tool linear movement are combined. Special devices such as the D'Andrea® head and a new concept of milling tool and part rotation at the same time like the Mori Seiki® spinning tool.

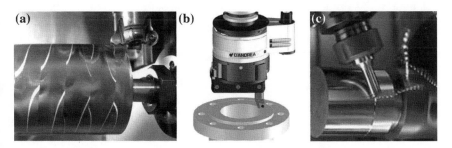

Fig. 7.3 a Turn-milling operation, **b** D'Andrea® head, **c** Spinning Tool Mori Seiki®

7.3 Propeller Blades Production Process

The new approach for the production of blades explained in this chapter consists of the following steps. The initial idea is to join all the production processes regarding the blade in the same machine. The multitasking concept of including as many operations in the same machine and set-up as possible considerably reduces possible manufacturing errors during the production process. Lead times are reduced as a direct consequence of not having to move the part from one machine to another. The same set-up allows maintaining the reference and possible errors due to changes in the part reference are also reduced.

Initially, the part needs to be turn-milled as to obtain a part near to net shape by roughing operations. After that, turn-ball-end milling for final part contouring is applied. In some cases, the use of balanced milling or pinch milling strategies leads to faster cycle times and balanced cutting forces thank to the appliance of two simultaneous operations with two simultaneous tools. Finally, the part is being burnished to improve the final roughness and fatigue endurance.

7.3.1 Turn-Milling

Turn-milling is a relatively new emerging concept inside the manufacturing technologies in 4- and 5-axis multitasking machines. To be able to carry out this operation, it is necessary to combine both the rotary movements of the tool and the part. This process, depending on the axis spatial position, can be classified into coaxial or orthogonal turn-milling [8]. It is named coaxial turn-milling when there is parallelism between the tool rotary axis and the part rotary axis. On the contrary, the process is named orthogonal when the tool rotary axis is perpendicular to the part rotary axis.

7.3.1.1 Coaxial Turn-Milling

In coaxial turn-milling operations, the tool part involved in the cutting process is the tool flank. A correct election of the milling tool is essential to develop the technique with satisfactory results; this is why contouring mills are the best ones for this kind of operations due to their geometry.

This type of turn-milling is the most appropriate one for external and internal milling when dealing with a revolution part. To perform this coaxial turn-milling, the rotation of the workpiece has to be quite slow in comparison with the tool rotary speed. As a consequence, there are a serial of advantages; the kinematic of the process causes shorter chips when the material to be machined is ductile. If it is necessary to machine rotary and slender parts, slower rotary part speeds avoid exciting the part at higher frequencies the same way it happens in conventional turning.

The machine tool needs to be able to interpolate three or four continuous axis to develop coaxial turn-milling operations. The kinematic of this technique is presented in the figure below. The tool needs to turn always in the same direction, as a consequence of the spatial location of the tool edges, special attention to the part rotation direction needs to be paid. The part rotation direction will determine whether the machining is up-milling or down-milling. Figure 7.4 shows up-milling operations; this is the strategy providing better results in relation to dimensional tolerance and surface finish.

Some special technique characteristics are as follows:

– By using the appropriate tool, internal machining can be carried out without previous holes.
– Non-cylindrical surfaces can be carried out as seen in the Fig. 7.4.
– Narrow channels can be machined.
– The technique is ideal for internal and external threads machined with thread mills.
– The principal disadvantage is the necessity of larger tool overhangs for deep cavities milling.

7.3.1.2 Orthogonal Turn-Milling

The process is named orthogonal when the tool rotary axis is perpendicular to the part rotary axis. This technique is only applied to external millings. There are several advantages related to this technique:

Fig. 7.4 Kinematic scheme of coaxial turn-milling operation and practical example

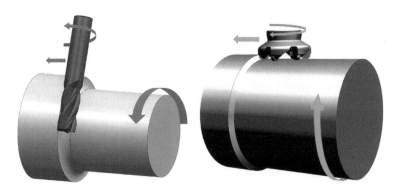

Fig. 7.5 Orthogonal turn-milling depending on the tool used

- Good results are obtained for thin walls machining. The cutting forces obtained are lower than the ones obtained in conventional turning.
- It is possible to use shorter tools overhang. The process becomes more stable when the chip removal rate is large.
- If the tool rotary speed is big enough and the optimal geometric parameters are used, the quality of the final surface obtained is very good even in comparison with grinding operations [9].
- The chip evacuation is much easier, and the machined area refrigeration is enhanced avoiding problems caused by excessive temperatures [10].
- The rotary part speed is so low that it is difficult to have machining problems due to centrifugal forces.

There is a variation of the orthogonal milling consisting of the location of the mill tangent to the part (Fig. 7.5). This technique has inspired several investigation works whose objective is to analyse and obtain machining parameters that will optimize the superficial roughness of the workpiece. In these cases, the working tools are finish mills and integral carbide mills [11, 12].

7.3.1.3 Theoretical Fundamentals of Orthogonal Turn-Milling

The initial point to define a turn-milling operation is the machining conditions imposed by the milling process; these conditions will determine the spindle speed, Eq. (7.1) and the linear tool speed, Eq. (7.2), as if it was a conventional milling process. The tool cannot move around the part, this is why the feed speed is given by the part to be machined. The calculation of the part rotation speed Eq. (7.3) is obtained from the previous relations Eq. (7.1) and Eq. (7.2) and the part diameter.

$$N = \frac{Vc \cdot 1000}{\varnothing_T \cdot \pi} \tag{7.1}$$

N (rpm) is the tool rotation speed, Vc(m/min) cutting speed and $\varnothing_T$(mm) total tool diameter.

$$V_f = Z_n \cdot f_z \cdot N \qquad (7.2)$$

V_f(mm/min) feed speed, Z_n number of tool edges, and f_z(mm/rev) feed per tooth.

$$\psi = \frac{V_f \cdot 360}{\varnothing_P \cdot \pi} \qquad (7.3)$$

ψ(°/min) part rotation speed and $\varnothing_p$ tool diameter. When programming the operation, it could be more interesting knowing the part rotation speed (rev/min), as seen in Eq. (7.4).

$$\varphi = \frac{V_f}{\varnothing_P \cdot \pi} \qquad (7.4)$$

For the axial feed calculation, it is necessary to determine the radial depth, ae(mm). Equation (7.5) shows the expression required for axial feed calculation.

$$V_a = a_e \cdot \varphi = a_e \cdot \frac{V_f}{\varnothing_P \cdot \pi} \qquad (7.5)$$

7.3.1.4 Tool Y-Axis Compensation

Orthogonal turn-milling operations can cause a characteristic half-moon print on the part surface. This effect is caused by the tool geometry and by the combined action of simultaneous tool and part rotation. Axial feed direction also shows a peculiar geometric form, depending in this case on the tool geometry and the tool Y-axis compensation value, called E_w(mm). Figures 7.6 and 7.7 show extreme cases.

One of these extreme cases, $E_w = 0 \% \varnothing_{ef}$, being $\varnothing_{ef}$ (mm) effective diameter of the cutting tool. This spatial position of the tool generates a convex surface in the machined surface, as seen in Fig. 7.6. This operation is recommended for slots finish operations, to be able to generate the fillet between the tool and the wall. The way the cut is made is similar to a turning operation (ramping), where the tool inserts machine with the lower part generating high axial forces.

The other extreme case is $E_w = 50 \% \varnothing_{ef}$. This spatial position of the tool generates a concave surface, Fig. 7.7. This displacement avoids the effect generated in the lower part of the tool, and consequently, axis cutting forces are reduced. This position avoids interference risks between the tool centre and the part whatever the axial depth is (ap).

Ew=0% Øef

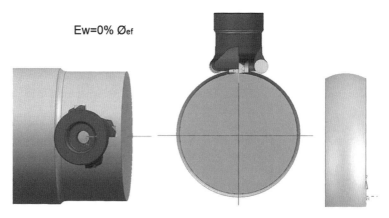

Fig. 7.6 Extreme cutting case Ew = 0 % Øef

Ew=50% Øef

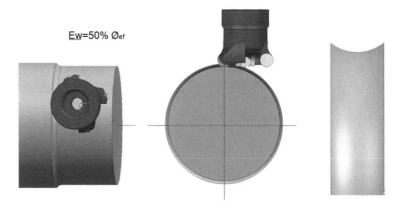

Fig. 7.7 Extreme cutting case Ew = 50 % Øef

7.3.2 Turn-Ball-End Milling

Once the part has gone through the roughing process, it is time for the finishing process. For blade, finishing turn-ball-end milling process is applied. The need of quality parts must constantly be balance with the need of cycle time reduction. This balance is even more noticeable in industries that require precise dimensional integrity on a range of slender workpieces that tend to deflect from tool pressure. One solution to this challenge is a multitasking machine, a machine equipped with two opposing turrets. These machines allow pinch (or balanced) machining, where two tools cut simultaneously on either side of the workpiece [13]. Some CAM softwares like ESPRIT® have already introduced this technology into their software. Simultaneously, cutting with two tools is not only faster but also equally important. The balanced cutting forces counteract the inclination of slender workpieces to deflect from the pressure of a single tool. Pinch machining has historically been used in turning applications, but recent developments in machine tool technology and CAM software are expanding pinch machining technology to milling as well. Pinch milling on a mill-turn can be better than a traditional machining centre for certain workpieces.

The main advantages of balanced milling or turn-milling operations are faster cycle times and balanced cutting forces.

7.3.2.1 Faster Cycle Times

When two tools cut the same profile simultaneously, with one tool above the axis of rotation and the other below, the feed rate for each tool can be doubled without affecting the tool load. Another technique is to start a cut with one tool, while the second tool waits until a fixed distance is reached. Then, the second tool follows the first to perform a second cut at a different depth, effectively performing two operations in the same pass.

7.3.2.2 Balanced Cutting Forces

Deflection is a serious problem when slender or long workpieces are machined with a single tool. Simultaneously, applying two tools on either side of the workpiece balances the cutting forces to counteract deflection from uneven tool pressure. Eliminating deflection is another factor that noticeably reduces cycle time while improving qualitative control.

7.3.2.3 Balanced Roughing

Balanced roughing function can be applied for pinch turning. The same cycle can be used for balanced finishing as well. A single balanced roughing operation can include roughing passes, finish passes, or both.

7.3.2.4 Simultaneous Roughing at Double the Feed Rate

The upper and lower tools rough the same profile at the same depth simultaneously (Fig. 7.8). Essentially, it consists of using two roughing operations at exactly the same time. This allows the feed rate to be doubled while maintaining the same cutter load. This is by far the fastest way to rough a profile.

Rough and re-rough at the same time, one tool leads and the second tool trails behind the other by a fixed distance. Both tools start at the same position, but the trailing tool waits until the leading tool reaches a user-defined distance before it starts cutting. In this case, the tools can cut at different depths to perform a rough and re-rough at the same time.

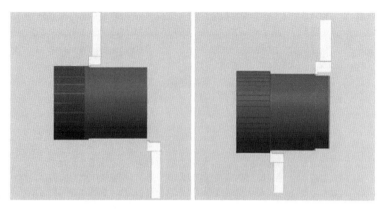

Initial roughing.   Simultaneous balance
Second tool ready for roughing   roughing

Fig. 7.8 Balance roughing

7.3.2.5 Balanced Finishing

When balanced roughing is used to create finishing passes only, contouring passes
are generated instead of roughing passes. A trailing distance is typically used for
balanced finishing. Different stock allowance can be set for each tool. The first tool
can perform the semi-finish at a set stock allowance, and the second tool performs
the finish cut to the net shape.

7.3.2.6 Inside and Outside Diameter Turning

Inside and outside diameter turning is shown in Fig. 7.9. The tool located on the
lower turret would crash into the turning tool on the upper turret if both operations
started at the same time. This is why synchronicity between operations is essential.

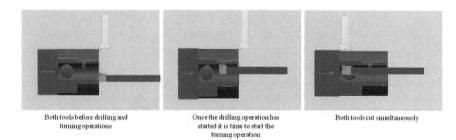

Both tools before drilling and   Once the drilling operation has   Both tools cut simultaneously
turning operations   started it is time to start the
                                         turning operation

Fig. 7.9 Simultaneous drilling and OD turning

Fig. 7.10 Pinch grooving

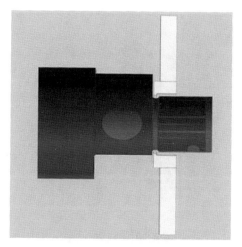

Simultaneous grooving

7.3.2.7 Pinch Grooving

Figure 7.10 shows separate grooving operations created on the upper and lower turrets. Both grooving operations have been synchronized at the start of each plunge or cut, so the two tools move together.

7.3.2.8 Two Different Tools in Different Orientations

Since any two turning operations can be performed simultaneously (within reason), it is also possible to find two different tools, operations and orientations simultaneously cutting.

These are just a few examples of how much time can be saved using the upper and lower turrets simultaneously.

7.3.2.9 Pinch Milling

Pinch milling (Fig. 7.11) is much like pinch turning except that the operations use live milling tools in the upper and lower turrets or spindles to machine opposite sides of the workpiece at the same time. Pinch milling and drilling is the most productive when both turrets allow off-centre Y-axis movement. However, even without a Y-axis, pinch slotting and drilling can easily be performed on the centre line.

It is important to synchronize the movement of the two milling tools. The milling operations can be created on the lower and upper turrets and then synchronized in the operation manager of the CAM software.

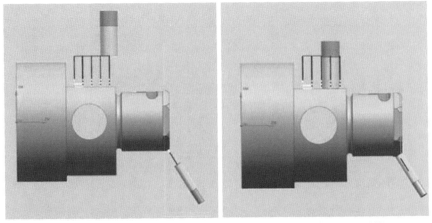

Initial milling operations   Face and threat milling

Fig. 7.11 Pinch milling

7.3.2.10 5-Axis Pinch Milling

Five-axis pinch milling (Fig. 7.12) technology represents a hard challenge in CAM programming because the upper and lower milling tools follow different contouring tool paths that must be carefully coordinated to synchronize the axis motions.

To be able to carry out 5-axis pinch milling strategies, a multitasking machine with a B-axis head and a lower turret that supports live tooling is required. Having opposing spindles is also recommended. The turning spindles provide the C-axis motion and provide balanced support at each end of the workpiece. This type of pinch milling is well suited for producing turbine blades since the blades are typically long, thin and prone to deflection. When the blade is supported between

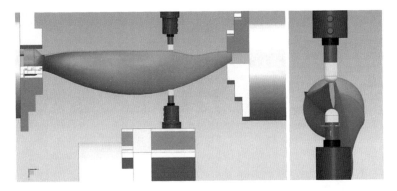

Fig. 7.12 (*Left*) 5-axis pinch milling frontal view of a propeller blade; (*Right*) 5-axis pinch milling lateral view of a propeller blade

opposing rotary spindles, the spindles can clamp and apply torque at both ends of the workpiece. Because the part can be held more rigidly and is subjected to far less deflection, allowable spindle load can be increased on both spindles. This configuration also permits the use of bull-nose end mills, which can take wider passes than a ball-nose without increasing cusp height. By keeping a bull-nose tool normal to the surface during simultaneous five-axis machining, the machine can maintain constant cutting speed. The result is faster metal removal and superior finish. Cutting with two tools lets shops crank up feed rates and maximize throughput with aggressive turning techniques.

7.3.3 Burnishing

7.3.3.1 Introduction

A great variety of products, for instance, die for automotive body, turbine blades and moulds, present complex surfaces, which can be of three types: geometrically ruled, sculptured (free-form) or spherical ones. These shapes are generally machined in multi-axis milling machines using ball-end milling tools, to be later on finished to mirror-like aspect if necessary, by hand polishing in a lot of cases. The final finishing and hand polishing processes could represent as much as 70 % of total machining time, being the reduction in this manufacturing step the key for a drastic reduction in lead times.

Mechanical surface treatments have been widely used to improve the physical–mechanical properties of metallic components [14]. As a consequence of plastic deformations, compressive residual stress states, work hardening, micro-structural alterations and a favourable roughness are produced, improving fatigue strength and wear resistance. Therefore, these surface treatments prevent crack initiation, retard propagation of small cracks, improve corrosion resistance, and even improve wear behaviour.

Ball burnishing is a rapid, simple and cost-effective mechanical surface treatment. This process is based on making small plastic deformations on part surfaces, which cause material displacement from the "peaks or ridges" to the "valleys or depressions" of the surface micro-irregularities. This mechanism is performed by a rolling element that moves over the tool paths on the surface, applying a regular compression force at the same time. Burnishing systems can be roller type [15], in this case the process is known as "rolling" or "roller burnishing", and ball type [16], whereby the term "ball burnishing" is used. Hydrostatic ball burnishing technology applied in 3- or 5-axis machines allows a finish surface within the quality of grinding (i.e. less than 1 μm Ra) [17]. Thus, ball burnishing can replace finishing processes such as grinding, shot peening or hand polishing. Deep ball burnishing commonly refers to a hydrostatic burnishing tool capable of supplying pressures of 20–30 MPa, introducing compressive residual stresses over 1 mm in depth. A variation of this technique is the so-called low-plasticity burnishing

(LPB) by Golden and Shepard [18]. LPB is a mechanical surface enhancement technology developed and patented at Lambda Technologies. This technique uses the minimal amount of plastic deformation needed to create the desired level of compressive stress. LPB is mainly applied to improve the fatigue properties of gas turbine engine components [19, 20]. This process causes four effects on the surface:

- Reduction in surface roughness in more than an order of magnitude. The final quality finishing is similar to grinding, even reaching mirror-like aspect in most of the cases.
- Generation of high compression residual stresses on the workpiece surface, which is beneficial for the fatigue behaviour of the component.
- Moreover, the absence of heat produced by this mechanical surface treatment prevents metallurgical changes on surfaces.
- Surface hardness increment between 30 and 60 % (HBN) considering the usual values of common steels (mild or medium carbon steels). However, the harder the steel, the more reduced is the burnishing effect.
- Dimensional tolerances are kept (< 0.01 mm), for example in burnished holes. Special tools for hole finishing is a typical application of spring-type burnishing devices, specially used in automotive parts.

7.3.3.2 Fundaments of Hydrostatic Ball Burnishing

The ball burnishing system used in this work, shown in Fig. 7.13, is based on a hydrostatic spring. The pressure is supplied by a hydraulic pump and is capable of pumping at a pressure of 40 MPa. The pump is placed next to the machine where is mounted the burnishing equipment. The most important element of the system is the ceramic ball, which supported hydrostatically thanks to the pressure from the pump.

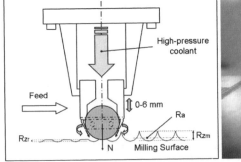 

Fig. 7.13 (*Left*) Principle of the hydrostatic ball burnishing. (*Right*) Detail of the ceramic ball on a flat surface (López de Lacalle et al. [21])

This ball smashes the roughness peaks derived from previous machining operations. The ceramic ball is placed on a head similar to the tip of a ball-type pen.

In the hydrostatic technology, the normal force only depends on the pump pressure. In addition, the ball head has a free movement of 6–8 mm, which helps to absorb errors in the tool path programming of the burnishing operation or allows some inaccuracy in the part profile, which facilitates the programming of the finishing process. In this stroke, the axial forces are kept constant, because force only depends on the pump pressure.

The burnishing technique can be used on cylindrical surfaces, flat surfaces, profiled surfaces, conical surfaces, zones of sharp section changes, etc. The main limitation of this process is basically geometric: Thus, the angle between the normal to the burnished surface and the tool axis must be between ± 28° to avoid lateral collisions. Within this range, the normal force remains approximately constant. However, beyond this slope, the tool or workpiece could be damaged by the ballholder contact against the burnished part. In the case of sculptured surfaces, part slope usually exceed this range in several zones; this is the main reason to work with multi-axis machines, that is, to avoid the risk of side collision of the ball head against a too steep surface.

A complete study of the effect of previous milling and burnishing parameters was presented in [21], and for the surface improvement of rotating shafts in [22], defining the window parameters for an optimal application of this technology on medium carbon steels. It concluded that using the optimum burnishing parameters—mainly burnishing pressure (P_b) and radial width of burnish (a_b))—it is possible to improve roughness quality around a 90 % and the surface hardness up to 60 % (in Brinell scale). The optimum parameters depend on the workpiece material and the quality of the previous machined surface.

In burnishing, process parameters have a great importance in achieving the optimum final values for roughness. For the same material, focusing on the variation of burnishing parameters, the finish quality depends on the following factors:

– Burnishing pressure and normal force. This value is limited by the maximum pressure of the external pressure unit. In our case, the hydraulic pump is capable of supplying coolant to a maximum pressure of 40 MPa. The burnishing force (F_b) can be estimated analytically by the fluid pressure (P_b) and the diameter of the ball (d). This relationship is given by the following equation [Eq. (7.6)]:

$$F_b = \frac{\pi}{4} \cdot d^2 \cdot P_b \qquad (7.6)$$

The usual difference with values obtained experimentally in the force measurement is 11 % [23]. This difference is due to small pressure losses along the fluid conductions and links.

– Workpiece material . The properties of the part material determine the surface deformation and therefore the final characteristics of roughness, hardness, etc.

- Previous machining roughness. The roughness obtained in a ball-end milling is directly related to the radial width of cut and the tool radius. Therefore, depending on the type of tool used for the previous machining and the programmed cutting conditions, different roughness is obtained. These roughness values affect the posterior burnishing process. Thus, a higher value of milling roughness will require a higher value of burnishing pressure to achieve an optimum surface quality.
- Feed rate. The maximum linear feed provided by the milling machine can be used for burnishing. In high-speed machines with ball-screw motion drives, this speed is usually between 10 and 20 m/min and somewhat higher when linear motors are used for the axis drives, for example 30 m/s.
- Radial width of burnish (a_b). This is a similar concept as the radial width of cut (a_e) for milling operations. The lower the value, the better surface finish is achieved. However, low values imply a longer processing time.
- Number of passes. Two or more passes improve surface quality, but normally a single pass is enough to increase significantly surface quality in comparison with previous milling.
- Burnishing direction. This direction is taken into account regarding the previous milling direction. There may be three different cases (a) perpendicular to milling tool paths, (b) same direction of milling tool paths and (c) at a certain angle with respect to the milling direction. In many cases, the milling tool and the burnishing tool have the same restrictions to act on the surfaces; therefore, there is usually only one possible way for milling and burnishing.
- Lubrication. Coolant forms an elasto-hydro-dynamic film between the ball and the workpiece surface, reducing friction and removing frictional heat. In our system, an oil emulsion to 8 % was used. Previous works [24] on the influence of the type of lubrication in this process have shown that the use of different types of lubricant barely affects the final roughness obtained.

7.3.3.3 Multi-Axis Ball Burnishing

A hemispherical surface demo-piece is presented in this section (see Fig. 7.14). This surface is simpler but has the same feature that free-form sculptured surfaces, that is, slopes are greater than ± 28° (the maximum slope to be climb by the burnishing ball), and so burnishing must be performed in a multi-axis machine in order to overcome this geometric limitation. The aim of this test workpiece was to study the feasibility of the process in sculptured surfaces using different strategies for burnishing. At the same time, spherical-type forms are common on hip implants, and a typical part made by five-axis ball-end milling.

The burnishing of this surface was performed with two different strategies: Continuous burnishing, using simultaneous 5-axis interpolation, and patch burnishing, doing it in different patches, so that becomes necessary to study the borderline between patches to verify that the surface quality is maintained. The

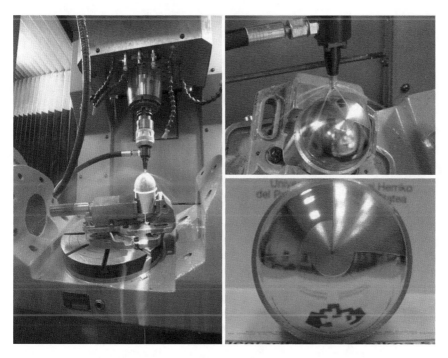

Fig. 7.14 Test piece. (*Left*) Burnishing system in 5-axis machine. (*Up*) Detail of the burnishing tool. (*Down*) Detail of the hemisphere burnished with different strategies (López de Lacalle et al. [21])

former is more complex to program but tool paths are continuous on the machined surface. The later is much easier to program in a multi-axis machine because rotary axes are blocked during the zig-zag of burnishing (in 3-axis).

The main advantage of continuous burnishing is the ability to access the complex zones on the part, achieving a significant improvement in the workpiece roughness evenly. In addition, the CAM programming of this type of strategy is relatively similar to the finish machining. On the other hand, there is one drawback of using this burnishing strategy: In 5-axis machines, usually rotating axis maximum speeds are lower than linear axis speeds; this implies that the burnishing feed rate is limited by the speed of rotary axes.

On the other hand, patch burnishing strategy allows the burnishing tool to produce patches that can be burnished in 3 axes, taking into consideration the maximum slope to climb by the burnishing ball (± 28° angles). To complete the total surface areas, the strategy would be 3 + 2 axis, burnishing using 3-axis, orientation of the part and burnishing in 3-axis again. In this way, it takes advantage in order to work at the maximum feed rate that allows the machine. Moreover, eliminates the need to interpolate the 5-axis simultaneously, reducing the possibility of errors and simplifying the CAM programming. The main disadvantage of this strategy is the heterogeneous finish obtained. The border line between passes has generated a roughness peak. The higher pressure is used, the higher the peak is.

7.3.4 Other Non-Traditional Processes Related to Blades

Complex parts and geometries like blades are subject to a serial of manufacturing process to be finished. Traditional machines development into multitasking machines has lump together different non-traditional processes such as laser cladding, and laser tempering or advanced coolant technologies application into the same machine.

7.3.4.1 Laser Cladding

If a part that has been damaged due to hard working conditions or if it has suffered from manufacturing defects, laser-cladding material deposition is a suitable technique to able in order to restore it. Laser-cladding process is one of the most relevant new processes in the industry due to the particular properties of the processed parts [25]. The main advantage is the possibility of obtaining high-quality material deposition on complex parts. Thus, laser cladding can be applied in the repair of high added-value and safety critical parts. This ability is especially useful for high-cost parts that present wears or local damage due to operating conditions. Different types of parts can be processed, such as housings, blades or even complete turbine rotors. Once the parts are repaired by laser cladding, they can be reassembled on the engine, reducing lead times.

The process consists of a method that utilizes a focused laser beam to melt the injected powder material as well as the surface of the substrate to produce a metallurgical bonded coating [26].

7.3.4.2 Laser Tempering

Specific zones of the blade, such as leading and output edges (Fig. 7.15), require a thermal treatment in order to modify the mechanical properties of material. For this purpose, different methods for heat treatment have been developed, laser tempering is an alternative technology for hardening treatment to generate in the critical area a greater hardness and thus conferring the treated surface with greater wear resistance.

Although the laser tempering treatment was one of the first applications of laser, the process has reached its full potential recently. The reason lies in the lack of knowledge about the process, the large number of alternatives for surface treatment that is on the market and the high cost of laser instrumentation. The heating

Fig. 7.15 Leading and output edges of the propeller blade

and cooling cycle of the material in a very short period of time makes the control of the process complex. Recent advances in the technology of the laser source, optics and software make the process to be more favourable compared to conventional competence processes.

From the economic point of view, the main advantages to emphasize the technique are that the thermally treated zone reaches a shallow depth; therefore, the energy required for heating is small; this makes the process having a high economic yield. As a consequence of the low amount of energy input, small deformations occur and/or dimensional changes of the workpiece after heat treatment; therefore, the piece needs not to be subjected to a finishing machining.

From the productivity and flexibility point of view, it should be noted that the energy source may be supplied by a wide range of lasers. The shape of the spot and the approach can be adapted to the characteristics of the process through the use of different lenses or mirrors. The extension, geometry or accessibility of the piece is not a problem because the tool used is "light". This technique is suitable to be incorporated in an automated production system.

7.3.4.3 Advanced Coolant Technologies

Lubri-coolants are commonly used in order to improve the productivity during machining. Emulsion oil lubricants and cutting oils have been used for years because of its proven cooling and lubricating power. Nowadays, new advanced coolant technologies are implemented due to strict regulations and ecological consciousness. These new technologies, such as cryogenic coolant, are more efficient, economical and environmentally sustainable. These techniques can be very useful to improve the productivity in propeller blades roughing operations, reducing tool wear, improving machining conditions and maintaining the surface integrity.

Cryogenic machining (Fig. 7.16) replaces conventional coolants by liquid nitrogen or carbon dioxide. This technology implies a perfect orientation through the tool tip and a great control of the flux parameters. The injection point is also important; the liquid nitrogen needs to reach both the rake face and the flank face of the tool. Cryogenic machining seems to be finding a place in this market due to its many advantages. However, it is a new technology by now little studied.

Liquid nitrogen at −196 °C is a colourless, non-toxic, inert gas, so it does not react with other materials and vaporizes quickly. Its advantages include increased cutting speed and material removal rates, with the consequent reduction in machining times. An alternative to the use of liquid nitrogen is the use of CO_2. Their lower price and ease of use made of CO_2 a feasible option for use as cryogenic cooling system. The solution is considered to be really advantageous for machining high strength materials and can generate numerous technological and economic benefits.

7.4 Examples

7.4.1 Propeller Blade

In this example (Table 7.1), the blade has been turn-milled making use of CA-PTO® system tools, the roughing operation lasted 6 min. Afterwards, it is subject to finishing turn-ball-end milling operation. This step of the manufacturing process lasts longer than the roughing process. Finally, some parts of the blade were burnished for better surface finish and also in order to avoid tensile stresses.

All the operations were virtually verified in order to avoid possible collisions or undesirable movements that will cause errors.

Table 7.1 Propeller blade manufacturing process

No of phase subphase operation	Phase description: phase (1000, 2000…) subphase (1100, 1200…) operation (1101, 1102…)	Fixtures Tools Control tools	Cutting conditions				Time (min)	Image Drawing of the part through the different stages of the process. Initial dimensions, fixtures, tools.
			Vc m/min	ap mm	ae mm	f_n/f_z mm/rev		
1	Turn-milling	CAPTO system Turning tool OD_80_L	600	5	75 %	0.2	6	
2	Turn-ball-end milling	CAPTO system Turning tool OD_80_L	800	0.2	25 %	0.1	13	
3	Burnishing	CAPTO system Turning tool OD_80_R	200	1	–	0.2	9	

7.5 Conclusions

The present work presents a non-traditional machining method for propeller blades manufacturing. It includes turn-milling, turn-ball-end milling and burnishing technologies, as well as complementary processes such as laser cladding, laser tempering and advanced coolant technologies for that could also be applied to blades.

Acknowledgments Thanks are addressed to the Department of Universities and Research of Industry of the Basque Government. The authors are grateful for funds of the UPV-EHU (UFI 11/29). Also special thanks are addressed to the ETORTEK PRO-FUTURE II and INNPACTO DESAFIO project.

References

1. Choi BK, Jerard RB (1998) Sculptured surface machining: theory and applications, Kluwer academic Publishers
2. Tönshoff HK, Gey C, Rackow N (2001) Flank milling optimization—the flamingo project. Air Space Europe 3:60–63
3. Bedi S, Mann S, Menzel C (2003) Flank milling with flat end milling cutters. Comput Aided Des 35:293–300
4. Li C, Bedi S, Mann S (2006) Flank milling of ruled surface with conical tools-an optimization approach. The Int J Adv Manufact Technol 29:1115–1124
5. http://www.free-online-private-pilot-ground-school.com/propeller-aerodynamics.html (2012)
6. http://www.thaitechnics.com/propeller/prop_intro.html (2012)
7. López de Lacalle LN, Lamikiz A (2009) Machine tools for high performance machining. Springer, London
8. Shaw MC, Smith PA, Cook NH (1952) The rotary cutting tool. ASME 74:1065–1076
9. Schulz H, Kneisel T (1994) Turn-milling of hardened steel and alternative to turning. CIRP 43:93–96
10. Chen P (1992) Cutting temperature and forces in machining of high performance materials with self-propelled rotary tool. JSME Int J Ser III 35:180–185
11. Choudhury SK, Bajpai JB (2005) Investigation in orthogonal turn-milling towards better surface finish. J Mater Process Technol 170:487–493
12. Savas V, Ozay C (2008) The optimization of the surface roughness in the process of tangential turn-milling using genetic algorithm. Int J Adv Manuf Technol 37:335–340
13. In a Pinch: using 2 tools to balance cutting forces www.dptechnology.com (2012)
14. Maawad E, Brokmeier H-G, Wagner L, Sano Y, Genzel Ch (2011) Investigation on the surface and near-surface characteristics of Ti–2.5Cu after various mechanical surface treatments. Surf Coat Technol 205:3644–3650
15. Maximov JT, Kuzmanov TV, Duncheva GV, Ganev N (2009) Spherical motion burnishing implemented on lathes. Int J Mach Tools Manuf 49:824–831
16. López de Lacalle LN, Lamikiz A, Muñoa J, Sánchez JA (2005) Quality improvement of ball-end sculptured surfaces by ball burnishing. Int J Mach Tools Manuf 45:1659–1668
17. Rodríguez A, López de Lacalle LN, Celaya A, Fernández A, Lamikiz A (2011) Ball burnishing application for finishing sculptured surfaces in multi-axis machines. Int J Mechatron Manuf Syst 4–3(4):220–237
18. Golden PJ, Shepard MJ (2007) Life prediction of fretting fatigue with advanced surface treatments. Mater Sci Eng 15–22:468–470

19. Prevéy P, Cammett J (2004) The influence of surface enhancement by low plasticity burnishing on the corrosion fatigue performance of AA7075-T6. Int J Fatigue 26(9):975–982
20. Prevéy P, Ravindranath R, Shepard M, Gabb T (2003) case studies of fatigue life improvement using low plasticity burnishing in gas turbine engine applications. In: Proceedings of ASME Turbo Expo, Atlanta, GA, pp 16–19
21. López de Lacalle LN, Rodriguez A, Lamikiz A, Celaya A, Alberdi R (2011) Five-axis Machining and Burnishing of Complex Parts for the improvement of surface roughness. Mater Manuf Process 26:997–1003
22. Rodríguez A, López de Lacalle LN, Celaya A, Lamikiz A, Albizuri J (2012) Surface improvement of shafts by the deep ball-burnishing technique. Surf Coat Technol 206:2817–2824
23. Rottger K (2002) Walzen hartgedrehter oberflaechen. PhD dissertation, WZL
24. Baskohov GF, Karpov NF (1973) How the viscosity of the coolant affects the burnishing process. Machine Tooling 44:61–62
25. Tabernero I, Lamikiz A, Martínez S, Ukar E, Figueras J (2011) Evaluation of the mechanical properties of Inconel 718 components built by laser cladding. Int J Mach Tools Manuf 51(6):465–470
26. Yangsheng L, Xue L (2010) An integrated software for laser cladding repair or worn-out components, Industrial Materials Institute, 561–570

Index

J. P. Davim (ed.), *Nontraditional Machining Processes*,
DOI: 10.1007/978-1-4471-5179-1, © Springer-Verlag London 2013

Printed by Printforce, the Netherlands